Show What You Know on the

MSP

**Preparation for the Measurements of Student Progress
Washington Comprehensive Assessment Program**

Name: Melanie Torres

Published by:

Show What You Know® Publishing
www.ShowWhatYouKnowPublishing.com
www.passthemsp.com

Distributed by:
Lorenz Educational Press, a Lorenz company
P.O. Box 802
Dayton, OH 45401-0802
www.LorenzEducationalPress.com

Copyright © 2009 by Show What You Know® Publishing
All rights reserved.

No part of this book, including interior design, cover design, and icons, may be reproduced or transmitted in any form, by any means (electronic, photocopying, recording, or otherwise).

MSP information was obtained from the Office of Superintendant of Public Instruction Web site, June 2009.

Printed in the United States of America

ISBN: 978-1-5923-0342-7

Limit of Liability/Disclaimer of Warranty: The authors and publishers have used their best efforts in preparing this book. Show What You Know® Publishing and the authors make no representations or warranties with respect to the contents of this book and specifically disclaim any implied warranties and shall in no event be liable for any loss of any kind including but not limited to special, incidental, consequential, or other damages.

Acknowledgements

Show What You Know® Publishing acknowledges the following for their efforts in making this assessment material available for Washington students, parents, and teachers:

Cindi Englefield, President/Publisher
Eloise Boehm-Sasala, Vice President/Managing Editor
Christine Filippetti, Production Editor
Jill Borish, Production Editor
Jennifer Harney, Editor/Illustrator

About the Contributors

The content of this book was written BY teachers FOR teachers and students and was designed specifically for the Measurements of Student Progress (MSP) for Grade 6. Contributions to the Reading and Mathematics sections of this book were also made by the educational publishing staff at Show What You Know® Publishing. Dr. Jolie S. Brams, a clinical child and family psychologist, is the contributing author of the Test Anxiety and Test-Taking Strategies chapters of this book. Without the contributions of these people, this book would not be possible.

Table of Contents

Introduction v

Test Anxiety 1

Test-Taking Strategies 11

Reading 21
    Introduction 21
    About the Reading MSP 22
        Item Distribution 22
        Scoring 23
    Glossary of Reading Terms 25
    Reading Practice Tutorial 31
    Reading Assessment One 49
    Reading Assessment Two 81

Mathematics 111
    Introduction 111
    About the Mathematics MSP 112
        Item Distribution 112
        Scoring 112
    Glossary of Mathematics Terms 113
    Glossary of Mathematics Illustrations 121
    Mathematics Practice Tutorial 127
    Mathematics Assessment 167

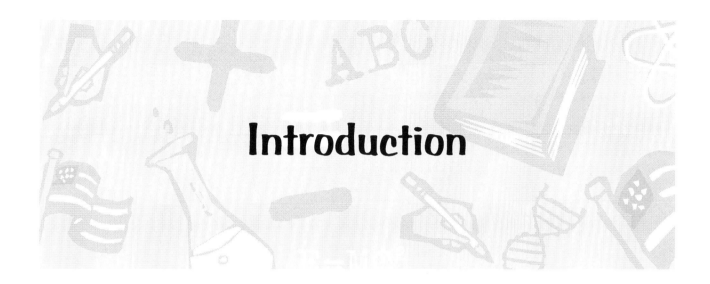

Dear Student:

This *Show What You Know® on the MSP for Grade 6, Student Workbook* was created to give you lots of practice in preparation for the Measurements of Student Progress (MSP) in Reading and Mathematics.

The first two chapters in this workbook—Test Anxiety and Test-Taking Strategies—were written especially for sixth-grade students. Test Anxiety offers advice on how to get rid of the bad feelings you may have about tests. The Test-Taking Strategies chapter gives you examples of the kinds of questions you will see on the MSP, such as multiple choice, short answer, and extended response and includes helpful tips on how to answer these questions correctly so you can succeed on the MSP.

The next two chapters of this Student Workbook help you prepare for the Reading and Mathematics MSP.
- The Reading chapter includes a Reading Practice Tutorial, two full-length Reading Assessments, and a Reading Glossary of words that will help you show what you know on the MSP.
- The Mathematics chapter includes a Mathematics Practice Tutorial, a full-length Mathematics Assessment, a Glossary of Mathematics Terms, and a Glossary of Mathematical Illustrations that will help you show what you know on the MSP.

This Student Workbook will help you become familiar with the look and feel of the MSP and will provide a chance to practice your test-taking skills to show what you know.

Good luck on the MSP!

This page intentionally left blank.

Test Anxiety

What is Test Anxiety?

Test anxiety is just a fancy name for feeling nervous about tests. Everyone knows what it is like to be nervous. Feeling nervous is not a good experience.

Many students have anxiety about taking tests, so if you are a test worrier, don't let it worry you. Most likely, many of your fellow students and friends also have fearful feelings about tests but do not share these feelings with others. Sixth grade is a time when everyone wants to seem "grown up," and few sixth graders want to look weak or afraid in the eyes of their friends or their teachers. But not talking to others about anxiety only makes the situation worse. It makes you feel alone and also makes you wonder if there is something "wrong" with you. Be brave! Talk to your friends and teachers about test anxiety. You will feel better for sharing.

What Does It Feel Like to Have Test Anxiety?

Students who have test anxiety don't always feel the same way, but they always feel bad. Here are some ways that students feel when they are anxious about tests.

- **Students who have test anxiety rarely think good things about themselves.**
  They lack confidence in their abilities, and they are convinced they will do poorly on tests. Not only do they feel bad about themselves and their abilities, but they just can't keep negative thoughts out of their minds. They would probably make terrible detectives, because in spite of all the good things they could find out about themselves, they only think about what they can't do. And that's not the worst of it. Students with test anxiety also exaggerate. When they think of the smallest problem, it becomes a hundred times bigger, especially when they think about tests. They are very unforgiving of themselves. If they make a mistake, they always think the worst or exaggerate the situation. If they do poorly on a quiz, they never say, "Well, it's just a quiz, and I'll try better next time." Instead they think, "That test was terrible and I can only imagine how badly I'll do next week." For students with test anxiety, there is never a brighter day ahead. They don't think many good thoughts about themselves, and they certainly don't have a happy outlook on their lives.

- **Students who have test anxiety have poor "thinking habits."**
  Negative thinking is a habit just like any other habit. Some habits are good and some habits are bad, but negative thinking is probably the worst habit of all. A habit forms when you do something over and over again until it becomes so much a part of you that you don't think about it anymore. Students with test anxiety get into bad thinking habits. They develop negative ways of thinking about themselves and about schoolwork, especially about tests. They tend to make the worst out of situations and imagine all kinds of possibilities that probably will not happen. Their thoughts grow like a mushroom out of control. Besides having negative ideas about tests, they begin to have negative ideas about almost everything else in their lives. This is not a good way of thinking because the more negative they feel about themselves, the worse they do in school, and bad grades make them feel even worse about themselves. What a mess. Students who have constant negative thoughts about themselves and schoolwork probably have test anxiety.

- **Students who have test anxiety may feel physically uncomfortable or even ill.**
It is important to know that your mind and body are connected. What goes on in your mind can change how your body feels, and how your body feels can influence what goes on in your thinking. When students have test anxiety, their thoughts might cause them to have physical symptoms which include a fast heartbeat, butterflies in the stomach, headaches, and all sorts of other physical problems. Some kids become so ill they end up going to the doctor because they believe they are truly sick. Some students miss a lot of school due to anxiety, but they aren't really ill. Instead, their thoughts are controlling their bodies in a negative way. Some anxious students do not realize that what they are feeling is anxiety. They miss many days of school, not because they are lazy or neglectful, but because they believe they truly are not feeling well. Unfortunately, the more school they miss, the more behind they are and the more nervous they feel. Students who suffer from test anxiety probably feel even worse on test days. Their uncomfortable physical feelings will make them either avoid the test completely or feel so bad during the test that they do poorly. Guess what happens then. They feel even worse about themselves, become more anxious, and the cycle goes on and on.

- **Students who have test anxiety "freak out" and want to escape.**
Many students feel so bad when they are anxious that they will do anything to avoid that feeling. For most students, this means running away from problems, especially tests. Some students try to get away from tests by missing school. This does not solve any problems; the more a student is away from school, the harder schoolwork is, and the worse he or she feels. Some students worry about being worried. It may sound silly, but they are worried that they are going to freak out, and guess what happens . . . they do. They are so terrified that they will have uncontrollable anxious feelings that they actually get anxious feelings when thinking about this problem. For many students, anxiety is such a bad feeling that they will do anything not to feel anxious, even if it means failing tests or school. Although they know this will cause them problems in the future, their anxiety is so overwhelming they would rather avoid anxiety now and fail later. Unfortunately, this is usually what happens.

- **Students who have test anxiety do not show what they know on tests.**
Students who have test anxiety do not make good decisions on tests. Instead of focusing their thoughts, planning out their answers, and using what they know, students find themselves "blanking out." They stare at the paper, and no answer is there. They become "stuck" and cannot move on. Some students come up with the wrong answers because their anxiety gets in the way of reading directions carefully and thinking about answers thoughtfully. Their minds are running in a hundred different ways and none of those ways seem to be getting them anywhere. They forget to use what they know, and they also forget to use study skills that can help students do their best. When students are so worried that they cannot make good decisions and use all of the talents they have, it is called test anxiety.

Are You One of These "Test-Anxious" Sixth Graders?

As you have seen, students with test anxiety have negative thoughts about themselves, often feel anxious to the point of being ill, freak out and want to escape, and rarely show what they know on tests. Do any of the following kids remind you of yourself?

Stay-Away Stephanie

Stephanie's thoughts tell her it is better to stay away from challenges, especially tests. Stephanie is a good girl, but she is always in trouble at school for avoiding tests. Sometimes, she really feels ill and begs her mom to allow her to stay home on test days. At other times, Stephanie does anything to avoid school, refusing to get up in the morning or to leave the house to catch the bus. Stephanie truly believes there is nothing worse than taking a test. She is so overwhelmed with anxiety that she forgets about the problems that will happen when she stays away from her responsibilities. Unfortunately, the more she stays away, the worse the situation becomes. Stay-Away Stephanie feels less nervous when she doesn't face a test, but she never learns to face her fears.

Worried Wendy

Wendy is the type of sixth grader who always expects the worst thing to happen. She has many negative thoughts. Even when situations have turned out to be OK, Wendy focuses on the few bad things that happened. She exaggerates negative events and forgets about everything good. Her mind races a mile a minute with all sorts of thoughts and ideas about tests. The more she thinks, the worse she feels, and her problems become unbelievably huge. Instead of just worrying about a couple of difficult questions on a test, she finds herself thinking about failing the whole test, being made fun of by her friends, being grounded by her parents, and never going to college. She completely forgets that her parents would never be so strict, that her friends like her for many more reasons than her test grades, and that she has all sorts of career choices ahead of her. No one is going to hold it against her if she performed poorly on a test. It is not going to ruin her life. However, Wendy believes all of that would happen. Her negative thoughts get in the way of thinking anything positive.

Critical Chris

Chris is the type of sixth grader who spends all of his time putting himself down. No matter what happens, he always feels he has been a failure. While some people hold grudges against others, Chris holds grudges against himself. No matter what little mistakes he makes, he can never forget them. Chris has had many good things happen to him in his life, and he has been successful many times. Unfortunately, Chris forgets all the good and only remembers the bad. Because he doesn't appreciate himself, Chris has test anxiety.

Victim Vince

Most sixth graders find it is important to take responsibility for their actions. It helps them understand that adulthood is just around the corner, and that they are smarter and more able than they ever thought they were. However, Vince is not like this. He can't take responsibility for himself at all. He thinks everything is someone else's fault and constantly complains about friends, parents, schoolwork, and especially tests. He tells himself, "They make those tests too hard." He sees the teachers as unfair, and he thinks life is generally against him. Vince does not feel there is anything he can do to help his situation, and there is little he thinks he can do to help himself with tests. Because he does not try to learn test-taking skills or to understand why he is afraid, he continues to feel hopeless and angry. Not surprisingly, he does poorly on tests, which only makes his thoughts about the world around him worse.

Perfect Pat

Everyone knows that there is more homework and responsibility in sixth grade than in previous grades. Everyone in the sixth grade needs to try his or her best, but no one should try as much as Pat does. All Pat does is worry. No matter what she does, it's never good enough. She will write book reports over and over and study for tests until she is exhausted. Trying hard is fine, but no matter what Pat does, she feels she has never done enough. Because she never accomplishes what she sets out to do (that would be impossible.), she worries all the time. Her anxiety level gets higher and higher. The more anxious she becomes, the worse she does on tests. This just makes her study and worry more. What a terrible situation!

How Do I Handle Test Anxiety?

Test anxiety is a very powerful feeling that convinces students they are weak and helpless. Feelings of test anxiety can be so powerful it seems there is nothing you can do to stop them. Anxiety seems to take over your mind and body and leaves you feeling like you are going to lose the test anxiety battle for sure.

The good news is that there are many simple things you can do to win the battle over test anxiety. If you can learn these skills in the sixth grade, you are on the road to success in school and for all other challenges in your life.

- **Change the way you think.**
  Most of us don't "think about how we think." We just go along thinking our thoughts and never really considering whether they are helpful or not helpful or if they are right or wrong. We rarely realize how much the way we think has to do with how well we get along in life. Our thoughts can influence how we feel about ourselves, how we get along with other people, how well we do in school, and how we perform on tests.

- **The Soda Pop Test.**
  Most sixth graders have heard a parent or teacher tell them, "There is more than one side to any story." One student reported that his grandfather used to say, "There's more than one way to paint a fence." Have you ever considered how you think about different situations? Most situations can be looked at in many ways, both good and bad.

Take a can of soda pop and put it on your desk or dresser at home. Get out a piece of paper and a pen or a pencil. Now, draw a line down the middle of the paper. On one side, put a heading: "All the bad things about this can of soda pop." On the other side put another heading: "All the good things about this can of soda pop." If you think about that can of soda pop, you might come up with the following chart.

All the bad things about this can of soda pop	All the good things about this can of soda pop
Not an attractive color	Easy-to-read lettering
It's getting warm	Nice to have something to drink
Not much in the can	Inexpensive
Has a lot of sugar	Recyclable aluminum cans

Look how easy it is to write down good things or bad things about a silly can of soda pop. That can of soda pop is not really good or bad, it's just a can of soda pop, but we can either look at it in a positive way or we can think about everything negative that comes to our minds. Doesn't the same thing hold true for tests? Tests are not good or bad in themselves. Tests are

just a way to challenge us and see what we know. Challenges can be stressful, but they can also be rewarding. Studying for tests can be boring and can take up a lot of free time, but we can also learn a lot and feel great about ourselves when we study. The way you think about tests will help determine how you do in a test-taking situation. Most importantly, how you feel about tests is related to your level of anxiety about test taking. Students who have negative thoughts and feelings about tests become anxious. Students who think positively are less anxious. To reduce test anxiety, try thinking about tests and testing situations using a positive frame of mind.

- **All or Nothing Thinking.**
Nothing is ever as simple as it seems. Sometimes we convince ourselves something is going to be "awful" or "wonderful." Rarely does it turn out that way.

Trouble comes along when students think tests are going to be an "awful" experience. If you dread something happening, it is only going to make things worse. Also, you may be wrong. Nothing is as terrible as it seems. All the negative thoughts you have about the upcoming test cannot possibly be true. Thinking something is "awful" or "terrible" and nothing else only leads to trouble and failure. The more negative you feel about something, the worse things turn out.

Very few things are "all good" or "all bad." This is especially true for tests. Recognizing the "bad" parts of tests can help you be successful. For example, the fact that you need to study for tests, to pay attention while you are taking tests, and to understand there are probably many more fun things to do in school than take tests are all "true" thoughts. "Good" thoughts are just as true, including the good feelings one gets from studying and the chance that you might do well. Having "all or nothing" thinking is going to get you nowhere. Successful and happy students know some experiences are better than others, but they try to look at a situation from all sides.

- **Mind Reading.**
Some students believe they can read the minds of their parents and teachers. They assume if they do poorly on the MSP, everyone will think they are "dumb" or "lazy." The more their minds create all the terrible things that people may say about them, the more anxious they get. This just increases anxiety and definitely does not help students do well on tests.

- **Catastrophizing.**
  When people catastrophize, they make everything a catastrophe. A catastrophe is a disaster. It is when something terrible happens. When a student catastrophizes, his or her mind goes on and on creating terrible scenes of disasters. If someone put all these ideas into a movie script, the writer might be rich.

  The MSP is an important part of a sixth-grader's school year. It is a test that helps the student, the teacher, and the school. However, a sixth-grade student is much more than just his or her score on the MSP. Each student is an individual who has his or her own great personality, talents, and other successes in school. If what people catastrophized about was really true, the whole world would be a terrible mess. Imagine if your mother cooked a dinner that didn't turn out quite right. This might mean everyone has to go out for fast food, but you wouldn't love your mother any less. It would be catastrophizing if your mother said, "Now that I burned the dinner, none of my kids will love me. They will probably just want to move out as quickly as they can, and my life will be ruined." Catastrophizing about the MSP is just as bad. Thinking that this test is going to be the worst experience of your life and that your future will be ruined will not help you feel comfortable when preparing for and taking the test.

- **Making "Should" Statements.**
  Students make themselves anxious when they think they "should" do everything. They feel they "should" be as smart as everyone else, "should" study more, and "should" not feel anxious about tests. All these thoughts are pretty ridiculous. You can't always be as smart as the next person, and you do not have to study until you drop to do well on tests. Instead of kicking yourself for not being perfect, it is better to think about all the good things you have done in your life. This will help you do better on tests and be happier in your life by reducing your anxiety.

How Do I Replace Worried Thoughts with Positive Ones?

As we have learned, there are all kinds of thoughts that make us anxious, such as feeling we "should" do everything, thinking we can read peoples' minds, catastrophizing, and thinking only bad thoughts about a situation. Learning how to stop these types of thoughts is very important. Understanding your thoughts and doing something about them help control test anxiety.

People who are worried or anxious can become happier when thinking positive thoughts. Even when situations are scary, such as a visit to the dentist, "positive imagery" is helpful. "Positive imagery" means thinking good thoughts to keep from thinking anxious thoughts. Positive and negative thoughts do not go together. If you are thinking something positive, it is almost impossible to think of something negative. Keep this in mind when test anxiety starts to become a bother.

Try these ideas the next time you find yourself becoming anxious.

- **Thoughts of Success.**
  Thinking "I can do it" thoughts can chase away thoughts of failure. Imagine times you were successful, such as when you performed well in a dance recital or figured out a complicated brain teaser. These are good things to think about. Telling yourself you have been successful in the past and can be successful in the future will chase away thoughts of anxiety.

- **Relaxing Thoughts.**
  Some people find that thinking calming or relaxing thoughts is helpful. Picturing a time in which you felt comfortable and happy can lessen your anxious feelings. Imagine yourself playing a baseball game, running through a park, or eating an ice cream cone; these are all positive thoughts that may get in the way of anxious ones. Some students find that listening to music on the morning of a test is helpful. It probably doesn't matter what music you listen to, as long as it makes you feel good about yourself, confident, and relaxed.

  Just as you can calm your mind, it is also important for you to relax your body. Practice relaxing your body. When students have test anxiety, their muscles become stiff. In fact, the whole body becomes tense. Taking deep breaths before a test and letting them out slowly as well as relaxing muscles in your body are all very helpful ways to feel less anxious. Your school counselors will probably have more ideas about relaxation. You may find that relaxation doesn't just help you on tests, but is helpful for other challenging situations and for feeling healthy overall.

- **Don't Let Yourself Feel Alone.**
  Everyone feels more anxious when they feel alone and separate from others. Talking to your friends, parents, and teachers about your feelings helps. Feeling anxious about tests does not mean there is something wrong with you. You will be surprised to find that many of your friends and fellow students also feel anxious about tests. You may be even more surprised to learn your parents and teachers have also had test anxiety. They know what you are going through and are there to support you.

- **Take Care of Yourself.**
  Everyone is busy. Many sixth graders are involved in all sorts of activities, including sports, music, and helping around the house. Often, you are so busy you forget to eat breakfast or you don't get enough sleep. Eating and sleeping right are important, especially before a test like the MSP. If you are not a big breakfast eater, try to find something that you like to eat and get in the habit of eating breakfast. When you do not eat right, you may feel shaky and have a hard time concentrating, and your anxiety can increase. Being tired does not help either. Try to get in the habit of going to bed at a good time every night (especially the night before a test) so you can feel fresh, rested, and confident for the MSP.

- **Practice Your Test-Taking Success.**
  People who have accomplished incredibly difficult goals have used their imaginations to help them achieve success. They thought about what they would do step by step to be successful.

  You can do the same. Think about yourself on the morning of the test. Imagine telling yourself positive thoughts and eating a good breakfast. Think about arriving at school and feeling confident that you will do fine on the test. Imagine closing your eyes before the test, breathing deeply, relaxing, and remembering all the study skills you have learned. The more you program your mind to think in a successful and positive way, the better off you will be.

- **Learn to Use Study Skills.**
  The next chapter in this book will help you learn test-taking strategies. The more you know about taking tests successfully, the calmer you will feel. Knowledge is power. Practice test-taking strategies to reduce your test anxiety.

- **Congratulate Yourself During the Test.**
  Instead of thinking, "I've only done five problems and I've got eight pages to go," or "I knew three answers were right but one mixed me up," reward yourself for what you have done. Tell yourself, "I got some answers right so far, so I bet I can do more." After all, if you don't compliment yourself, who will?

Conclusion

You are not alone if you feel stressed about tests. It is probably good to feel a little anxious, because it motivates you to do well. However, if you feel very anxious about tests, then reading, re-reading, and practicing the suggestions in this chapter will help you "tackle your test anxiety."

Test-Taking Strategies

All Students Can Do Their Best on Tests!

Most students want to do their best on tests. Tests are one important way for teachers to know how well students are doing and for students to understand how much progress they are making in their studies. Tests like the MSP help schools measure how well students are learning so teachers and principals can make their schools even better. Students can do the best job possible in "showing what they know" by learning how to be good test takers.

It's just not possible to do a good job without the right tools. Test-taking strategies are tools to help you perform well on tests. Everyone needs good tools and strategies when facing a problem. If you do not have these, even the smartest or most talented person will do poorly. Think about people who are "wizards" at fixing cars and trucks. Your family's car "dies" in the middle of the road. The situation looks pretty hopeless. How are you ever going to get to that basketball game tomorrow if your parent's car is a mechanical mess? Suddenly, "magic" happens. The mechanic at the repair shop calls your parents and tells them the car is ready, just a day after it broke down. How did this happen? It happened because the auto-repair mechanic had a great deal of knowledge about cars. Most importantly, he had the right tools and strategies to fix the car. He knew how to look at the problem, and when he figured out what to do, he had some special gadgets to get the job done. You also can find special ways that will help you be a successful test taker.

Tools You Can Use on the MSP and Tests Throughout Your Life!

Be An "Active Learner."
You can't learn anything by being a "sponge." Just because you are sitting in a pool of learning (your classroom) does not mean you are going to learn anything just by being there. Instead, students learn when they actively think and participate during the school day. Students who are active learners pay attention to what is being said. They also constantly ask themselves and their teachers questions about the subject. When able, they participate by making comments and joining discussions. Active learners enjoy school, learn more, feel good about themselves, and usually do better on tests. Remember the auto-repair mechanic? That person had a lot of knowledge about fixing cars. All the tools and strategies in the world will not help unless you have benefited from what your teachers have tried to share.

Being an active learner takes time and practice. If you are the type of student who is easily bored or frustrated, it is going to take some practice to use your classroom time differently. Ask yourself the following questions.

- Am I looking at the teacher?

- Do I pay attention to what is being said?

- Do I have any questions or ideas about what the teacher is saying?

- Do I listen to what my fellow students are saying and think about their ideas?

- Do I work with others to try to solve difficult problems?

- Do I look at the clock and wonder what time school will be over, or do I appreciate what is happening during the school day and how much I can learn?

- Do I try to think about how my schoolwork might be helpful to me now or in the future?

Although you do need special tools and strategies to do well on tests, the more you learn, the better chance you have of doing well on tests. Think about Kristen.

There was a young girl named Kristen,
Who was bored and wouldn't listen.
She didn't train
To use her smart brain
And never knew what she was missing!

Don't Depend on Luck.
Preparing for the MSP might feel stressful or boring at times, but it is an important part of learning how to show what you know and doing your best. Even the smartest student needs to spend time taking practice tests and listening to the advice of teachers about how to do well. Luck alone is not going to help you do well on the MSP or other tests. People who depend on luck do not take responsibility for themselves. Some people who believe in luck do not want to take the time and effort to do well. It is easier for them to say, "It's not my fault I did poorly. It's just not my lucky day." Some people just do not feel very good about their abilities. They get in the habit of saying, "Whatever happens will happen." They believe they can never do well no matter how much they practice or prepare. Students who feel they have no control over what happens to them usually have poor grades and do not feel very good about themselves.

Your performance on the MSP (and other tests) is not going to be controlled by luck. Instead, you can have a lot of control over how well you do in many areas of your life, including test taking. Don't be like Chuck.

There was a cool boy named Chuck,
Who thought taking tests was just luck.
He never prepared.
He said, "I'm not scared."
When his test score appears, he should duck!

Do Your Best Every Day.
Many students find sixth grade much different than other grades. Suddenly, the work seems really hard. Not only that, but your teachers are no longer treating you like a baby. That's good in some ways, because it gives you more freedom and responsibility, but there sure is a lot to learn. You might feel the same way about the MSP; you may feel you'll never be prepared. Many times when we are faced with new challenges, it is easy just to give up.

Students are surprised when they find that if they just set small goals for themselves, they can learn an amazing amount. If you learn just one new fact every day of the year, at the end of the year, you will know 365 new facts. You could use those to impress your friends and family. Now think about what would happen if you learned three new facts every day. At the end of the year, you would have learned 1,095 new facts. Soon you will be on your way to having a mind like an encyclopedia.

When you think about the MSP or any other academic challenge, try to focus on what you can learn step by step and day by day. You will be surprised how all of this learning adds up to make you one of the smartest sixth graders ever. Think about Ray.

There was a smart boy named Ray,
Who learned something new every day.
He was pretty impressed
With what his mind could possess.
His excellent scores were his pay!

Get to Know the MSP.

Most sixth graders are probably pretty used to riding in their parents' cars. They know how to make the air-conditioning cooler or warmer, how to change the radio stations, and how to adjust the volume on the radio. Think about being a passenger in a totally unfamiliar car. You might think, "What are all those buttons? How do I even turn on the air conditioner? How do I make the window go up and down?" Now, think about taking the MSP. The MSP is a test, but it may be different than some tests you have taken in the past. The more familiar you are with the types of questions on the MSP and how to record your answers, the better you will do. Working through the Reading and Mathematics chapters in this book will help you get to know the MSP. Becoming familiar with the MSP is a great test-taking tool. Think about Sue.

There was a kid named Sue,
Who thought her test looked new.
"I never saw this before!
How'd I get a bad score?"
If she practiced, she might have a clue!

Read Directions and Questions Carefully!

One of the worst mistakes a student can make on the MSP is to ignore directions or to read questions carelessly. By the time some students are in the sixth grade, they think they have heard every direction or question ever invented, and it is easy for them to "tune out" directions. Telling yourself, "These directions are just like other directions," or "I'm not really going to take time to read this question because I know what the question will be," are not good test-taking strategies. It is impossible to do well on the MSP without knowing what is being asked.

Reading directions and questions slowly, repeating them to yourself, and asking yourself if what you are reading makes sense are powerful test-taking strategies. Think about Fred.

There was a nice boy named Fred,
Who forgot almost all that he read.
The directions were easy,
But he said, "I don't need these."
He should have read them instead.

Know How to Fill in Those Answer Bubbles!

Most sixth graders have taken tests that ask them to fill in answer bubbles. You might be a very bright sixth grader, but you will never "show what you know" unless you fill in the answer bubbles correctly. Don't forget: a computer will be "reading" your multiple-choice question answers. If you do not fill in the answer bubble darkly or if you use a check mark or dot instead of a dark mark, your smart thinking will not be counted. Look at the examples given below.

Correct: ●   Incorrect: ✓ ✗ ● ◐ ◑

Practice filling in the answer bubbles here: ○ ○ ○ ○ ○ ○

Learning how to fill in answer bubbles takes practice, practice, and more practice. It may not be how you are used to answering multiple-choice questions, but it is the only way to give a right answer on the MSP. Think about Kay!

A stubborn girl named Kay
Liked to answer in her own way.
Her marked answer bubbles
Gave her all sorts of troubles.
Her test scores ruined her day!

Speeding Through the Test Doesn't Help.
Most students have more than enough time to read and answer all the questions on the MSP. There will always be some students who finish the test more quickly than others, but this does not mean the test was easier for them or their answers are correct. Whether you finish at a faster rate or at a slower rate than other students in your class is not important. As long as you take your time, are well prepared, concentrate on the test, and use some of the skills in this book, you should be able to do just fine. You will not get a better score just because you finish the test before everyone else. Speeding through a test item or through the whole MSP does not help you do well. In fact, students do their best when they work at a medium rate of speed, not too slow and not too fast. Students who work too slowly tend to get worried about their answers and sometimes change correct answers into incorrect ones. Students who work too fast often make careless mistakes, and many of them do not read directions or questions carefully. Think about Liz.

There was a sixth grader named Liz,
Who sped through her test like a whiz.
She thought she should race
At a very fast pace,
But it caused her to mess up her quiz.

Answer Every Question.
There is no reason that you should not attempt to answer every question you encounter on the MSP. Even if you don't know the answer, there are ways for you to increase your chances of choosing the correct response. Use the helpful strategies described below to help you answer every question to the best of your ability.

- **If you don't know the answer, guess.**

    Did you know that on the MSP there is no penalty for guessing? That is really good news. That means you have a one out of four chance of getting a multiple-choice question right, even if you just close your eyes and guess. That means that for every four questions you guess, you should get about 25% (1 out of 4) of the questions right. Guessing alone is not going to make you a star on the MSP, but leaving multiple-choice items blank is not going to help you either.

    Now comes the exciting part. If you can rule out one of the four answer choices, your chances of answering correctly are now one out of three. You can almost see your MSP score improving right before your eyes.

    Although it is always better to be prepared for the test and to study in school, we all have to guess at one time or another. Some of us do not like to guess because we are afraid of choosing the wrong answer, but on the MSP, it is better to guess than leave an answer blank. Think about Jess.

There was a smart girl named Jess,
Who thought it was useless to guess.
If a question was tough,
She just gave up.
This only added to her stress.

- **Use a "code" to help you make good guesses on the MSP.**

    Some students use a "code" to rate each answer when they feel they might have to guess. Using your pencil in the test booklet, you can mark the following codes next to each multiple-choice response so you can make the best possible guess. The codes are as follows:

(+)   Putting a "plus sign" by your answer means you are not sure if this answer is correct, but you think this answer is probably more correct than the others.

(?)   Putting a "question mark" by your answer means you are unsure if this is the correct answer, but you don't want to rule it out completely.

(−)   Putting a "minus sign" by your answer means you are pretty sure this is the wrong answer. You should then choose from the other answers to make an educated guess.

Remember, it is fine to write in your test booklet. Think about Dwight.

There was a smart kid named Dwight,
Who marked answers that looked to be right.
He'd put a plus sign
Or a dash or a line.
Now the whole world knows he is bright!

- **Use what you know to "power guess."**
Not everything you know was learned in a classroom. Part of what you know comes from just living your life. When you take the MSP, you should use everything you have learned in school, but you should also use your experiences outside the classroom to help you answer questions correctly. Using your "common sense," as well as other information you know, will help you do especially well on the MSP. Try to use what you know from the world around you to eliminate obviously wrong answers. If you can rule out just one answer that you are certain is not correct, you are going to greatly increase your chances of guessing another answer correctly. For example, if you are given a question in which you are asked to find the square footage of a home, and one of the answers seems very small, you might be able to count that answer out using your own experiences. Although the mathematics might be difficult for you, your common sense has eliminated one likely wrong answer. Think about Drew.

There was a boy named Drew,
Who forgot to use what he knew.
He had lots of knowledge.
He could have been in college!
But his right answers were few.

- **Do Not Get Stuck on One Question.**
One of the worst things you can do on the MSP is to get stuck on one question. The MSP gives you many chances to show all that you have learned. Not knowing the answer to one or two questions is not going to hurt your test results very much.

When you become stuck on a question, your mind plays tricks on you. You begin to think that you are a total failure, and your worries become greater and greater. This worrying gets in the way of your doing well on the rest of the test. Remember, very few students know all the answers on the MSP. If you are not sure of the answer after spending some time on it, mark it in your test booklet and come back to it later. When you to come back to that question later, you might find a new way of thinking. Sometimes, another question or answer later in the test will remind you of a possible answer to the question that had seemed difficult. If not, you can use your guessing strategies to solve the questions you are unsure of after you have answered all the questions you know. Also, when you move on from a troubling question and find you are able to answer other questions correctly, you will feel much better about yourself and you will feel calmer. This will help you have a better chance of succeeding on a question that made you feel "stuck." Think about Von.

There was a sweet girl named Von,
Who got stuck and just couldn't go on.
She'd sit there and stare,
But the answer wasn't there.
Before she knew it, all the time was gone.

- **Always, and This Means Always, Recheck Your Work.**
Everyone makes mistakes. People make the most mistakes when they feel a little worried or rushed. Checking your work is a very important part of doing your best on the MSP. This is particularly true in the Mathematics section, where careless mistakes can lead to a wrong answer, even when the student used the right steps. Going back and rechecking your work is very important. You can read a paragraph over again if there is something you do not understand or something that you forgot. In the Mathematics section, look at your calculations to make sure that you did not mistake one number for another and that you lined up your

calculations neatly and legibly. If any numbers seem messy or unreadable, you might want to recheck your calculations. If an answer does not seem to make sense, go back and reread the question, or recheck your work. Think about Jen.

There was a quick girl named Jen,
Who read stuff once and never again.
It would have been nice
If she'd reread it twice.
Her test scores would be better then!

- **Pay Attention to Yourself and Not Others.**
  It is easy to look around the room and wonder how friends are doing on the MSP. However, it is important to think about how you are using tools and strategies on the MSP. Don't become distracted by friends. You are going to waste a lot of time if you try to figure out what your friends are doing. Instead, use that time to "show what you know."

If it becomes hard for you to pay attention, give yourself a little break. If you feel you are getting a little tense or worried, or if a question seems tough, close your eyes for a second or two. Think positive thoughts about the MSP. Try to put negative thoughts out of your mind. You might want to stretch your arms or feet or move around a little to help you focus. Anything you may do to help pay better attention to the test is a great test-taking strategy. Think about Kirk.

There was a boy named Kirk,
Who thought of everything but his work.
He stared into the air
And squirmed in his chair.
When his test scores come, he won't look!

Reading

Introduction

The Reading Assessment of the Measurements of Student Progress (MSP) will measure how well you understand what you read. The Reading Assessment is based on the reading skills you have been taught in school through sixth grade. It is not meant to confuse or trick you but to allow you to have the best chance to show what you know.

The *Show What You Know® on the MSP for Grade 6, Student Workbook* includes a Reading Practice Tutorial that will help you practice your test-taking skills. Following the Reading Practice Tutorial, there are two full-length Reading Assessments. Both the Reading Practice Tutorial and the Reading Assessments have been created to model the Grade 6 Measurements of Student Progress for Reading.

Item Distribution-Reading · Show What You Know® on the MSP for Grade 6

About the Reading MSP

The Grade 6 Reading Assessment consists of six reading selections: two literary passages, two informational passages, and one paired passage, which could consist of informational/informational, informational/literary, or literary/literary. The Reading Assessment is given in one session.

For the Reading Assessment, you will read stories and other passages and answer some questions. There are three different types of questions. There are multiple-choice, short-answer, and extended-response questions. You may look back at the story or passage when you are answering the questions. However, you may not use resource materials during the Reading Assessment.

Item Distribution on the MSP for Grade 6 Reading

Text Types/ Strands	Number of Reading Selections	Number of Multiple Choice	Number of Short Answer	Number of Extended Response
Literary Comprehension Analysis	3	12 6 6	2–4 1–2 1–2	0 0–2 0–2
Informational Comprehension Analysis	3	12 6 6	2–4 1–2 1–2	0 0–1 0–1
Total	6	24	7	2

Scoring

On the MSP for Grade 6 Reading Assessment, each multiple-choice item is worth one point. Short-answer items will be scored on a scale of zero to two points and extended-response items will be scored on a scale of zero to four points. Responses are scored with emphasis on communication of ideas. Conventions of writing (sentence structure, word choice, usage, grammar, and mechanics) are generally disregarded unless they substantially interfere with communications.

Typical Distribution of Score Points by Item Type*

Type	Number of Items	Total Possible Points
Multiple Choice	24	24
Short Answer	7	14
Extended Response	2	8
Total	33	46

2009 testing information.

Scoring Rules for Short-Answer Items

Scoring rules for items that assess main ideas and details:
A **2-point** response shows thorough comprehension of the main idea and important details. It uses ample, relevant information from text(s) to support responses.

A **1-point** response shows partial comprehension of the main idea and important details (may grasp main idea but show difficulty distinguishing between important and unimportant details; may miss part of fundamental who/what/where/when/why). It attempts to use information from text(s) to support responses; support may be limited or irrelevant.

A **0-point** response shows little or no understanding of the passage main ideas and details.

Scoring rules for items that assess analysis, interpretation, and critical thinking about text:
A **2-point** response analyzes appropriate information and/or makes thoughtful connections between whole texts/parts of texts. It develops thoughtful interpretations of text. It uses sufficient, relevant evidence from text(s) to support claims.

A **1-point** response analyzes limited information and/or makes superficial connections between whole texts/parts of texts. It develops conventional or simplistic interpretations of text. It attempts to use evidence from text(s) to support claims; support may be limited or irrelevant.

A **0-point** response shows little or no understanding of the passage main ideas and details.

Scoring rules for items that assess summarizing and paraphrasing main ideas:
A **2-point** response shows thorough comprehension of main ideas.

A **1-point** response shows partial comprehension of main ideas.

A **0-point** response shows little or no understanding of the passage main ideas and details.

Scoring Rules for Extended-Response Items

Scoring rules for items that assess <u>analysis, interpretation, and critical thinking about text</u>:
A **4-point** response meets all relevant criteria. It thoroughly analyzes appropriate information and/or makes insightful connections between whole texts/parts of texts. It develops insightful interpretations of text. It uses ample, relevant evidence from text(s) to support claims.

A **3-point** response meets most relevant criteria. It analyzes appropriate information and/or makes thoughtful connections between whole texts/parts of texts. It develops thoughtful interpretations of text. It uses sufficient, relevant evidence from text(s) to support claims.

A **2-point** response meets some relevant criteria. It analyzes limited information and/or makes superficial connections between whole texts/parts of texts. It develops conventional or simplistic interpretations of text. It attempts to use evidence from text(s) to support claims; support may be limited or irrelevant.

A **1-point** response meets few relevant criteria. It shows difficulty analyzing information and/or makes weak connections between whole texts/parts of texts. It may not develop beyond literal interpretation of text. It uses little or no evidence to support claims.

A **0-point** response shows little or no understanding of the passage main ideas and details.

Glossary

affixes: Groups of syllables (i.e. prefixes, such as *anti–* or *post–*, and suffixes, such as *–ly* or *–ment*) which, when added to a word or a root, alters the meaning of the word.

alliteration: The repetition of the same sound, usually of a consonant, at the beginning of two or more words of a sentence or line of poetry (e.g., "Andrew Alligator always eats alphabet soup").

alliterative sentences: Repeating the same initial sound in two or more words of a sentence or line of poetry (e.g., Whitman's line, "all summer in the sound of the sea").

analogy: A comparison of two pairs that have the same relationship. The key is to discover the relationship between the first pair, so you can choose the correct second pair (e.g., part-to-whole, opposites).

analysis: Separation of a whole into its parts for individual study.

analyze: To compare in order to rank items by importance or to provide reasons. Identify the important parts that make up the whole and determine how the parts are related to one another.

anticipation guide: A flexible strategy used to activate students' thoughts and opinions about a topic and to link their prior knowledge to new material. For example, a series of teacher-generated statements about a topic that students respond to and discuss before reading.

antonyms: Words that mean the opposite (e.g., *light* is an antonym of *dark*).

assumptions: Statements or thoughts taken to be true without proof.

author's craft: Stylistic choices the author makes regarding such components as plot, characterization, structure, scenes, and dialogue to produce a desired effect.

author's perspective: The author's subjective view as reflected in his/her written expression.

author's purpose: The reason an author writes, such as to entertain, inform, or persuade.

author's style: The author's attitude as reflected in the format of the author's written expression.

author's tone: The author's attitude as reflected in the word choice of the author's written expression.

automaticity: The ability to recognize a word (or series of words) in text effortlessly and rapidly.

blend: In decoding, it is the reader's act of sounding out and then combining the sounds in a word to assist in the pronunciation.

common consonant sounds: Speech sounds made by obstructing air flow, causing audible friction in varying amounts. Common consonant sounds include: /b/, /k/, /d/, /f/, /g/, /h/, /j/, /l/, /m/, /n/, /p/, /kw/, /r/, /s/, /t/, /v/, /w/, /ks/, /y/, /z/.

common inflectional ending: A common suffix that changes the form or function of a word, but not its basic meaning, such as "–ed" in "sprayed," "–ing" in "gathering."

common sight words: Words that are immediately recognized as a whole and do not require word analysis for identification. These words usually have irregular spellings.

common vowel patterns: A vowel is the open sound. The mouth must be open to produce the sound of a vowel in a syllable. The most common vowel patterns are the sound/spellings that students encounter most frequently in text (e.g., ea, ee, oi, ow, ou, oo).

comprehension-monitoring strategies: Strategies used to monitor one's reading by being aware of what one does understand and what one does not understand. The reader's awareness determines which comprehension-repair strategies to apply.

comprehension-repair strategies: Strategies used by a reader to regain comprehension as a result of comprehension monitoring. These strategies include but are not limited to: re-reading, word recognition strategies, looking back, reading ahead, slowing down, paraphrasing by sections, using context, and taking notes. (Also referred to as "fix-up strategies.")

Glossary

comprehension strategies: A procedure or set of steps to follow in order to enhance text understanding (e.g., making inferences, predicting outcomes).

concepts of print: Insights about the ways in which print works. Basic concepts about print include: identification of a book's front and back covers and title page; directionality (knowledge that readers and writers move from left to right, top to bottom, front to back); spacing (distance used to separate words); recognition of letters and words; connection between spoken and written language; understanding of the function of capitalization and punctuation; sequencing and locating skills.

content area vocabulary: Vocabulary found in specific subject areas (e.g., "integer" in math and "pioneer" in social studies).

content/academic text: Text from literature, science, social studies, math, and other academic areas that students need to read to be academically successful in school.

content/academic vocabulary: Terms from literature, science, social studies, math, and other academic areas that students need to understand in order to be successful readers.

context: The social or cultural situation in which the spoken or written word occurs; also often used to refer to the material surrounding an unknown word.

context clues: Information from the surrounding text that helps identify a word or word group. Clues could be words, phrases, sentences, illustrations, syntax, typographic signals, definitions, examples, or restatements.

culturally relevant: Reading materials with which students in a classroom can identify or relate. Depending on the student cultural make-up in a classroom, relevant reading material can change from year to year.

decodable text: Reading materials that provide an intermediate step between words in isolation and authentic literature. Such texts are designed to give students an opportunity to learn to use their understanding of phonics in the course of reading connected text. Although decodable texts may contain sight words that have been previously taught, most words are wholly decodable on the basis of the letter-sound and spelling-sound correspondences taught and practiced in phonics lessons.

directionality: Understanding that materials printed in English progress from left to right and top to bottom.

electronic sources: Resources for gathering information such as the Internet, television, radio, CD-ROM encyclopedia, and so on.

elements of style: Word choice, voice, sentence structure, and sentence length.

environmental print: Any print found in the physical environment, such as street signs, billboards, labels, and business signs.

figurative language: Word images and figures of speech used to enrich language (e.g., simile, metaphor, personification).

fluency: Ability to read a text quickly with accuracy and expression; freedom from word-identification problems that might hinder comprehension in silent reading or the expression of ideas in oral reading; automaticity.

foreshadowing: A literary technique of giving clues about an event before it happens.

functional document: A technical document such as a business letter, computer manual, or trade publication that assists one in getting information in order to perform a task.

generalize: Taking what is known and using it to make an inference about the nature of similar text. Generalizations lead to transferable understandings that can be supported by fact. They describe the characteristics of classes or categories of persons, places, living and non-living things, and events.

Glossary

genres: Terms used to classify literary and informational works into categories (e.g., biography, mystery, historical fiction).

gist: The most central thought or idea in a text.

graphic features: Features that illustrate information in text such as graphs, charts, maps, diagrams, tables, etc.

graphic organizers: Organizers that provide a visual representation of facts and concepts from a text and their relationships within an organized frame. Valuable instructional tools used to show the order and completeness of a student's thought process graphically.

icons: Symbols on a computer screen that represent a certain function, command, or program on the computer's hard drive. When an icon is clicked on, some action is performed, such as opening or moving a file, making computing more user-friendly.

idiom: A word used in a special way that may be different from the literal meaning (e.g., "you drive me crazy" or "hit the deck").

independent level: The level at which the student reads fluently with excellent comprehension. The student demonstrates 95–100% comprehension of text.

infer: To understand something not directly stated in the text by using past experience and knowledge combined with the text.

inference: The reasoning involved in drawing a conclusion or making a logical judgment on the basis of indirect evidence and prior conclusions rather than direct evidence from the text.

inferred: Reached a specific conclusion using past experiences and knowledge combined with text evidence.

inflectional ending: A letter or group of letters which when added to the end of a word does not change its part of speech, but adjusts the word to fit the meaning of the sentence (e.g., girl, girls, jump, jumped, big, bigger).

informational/expository text: A form of written composition that has as its primary purpose explanation or the communication of details, facts, and discipline- or content-specific information (e.g., content area textbooks, encyclopedias, biographies).

instructional level: The level at which the student can make maximum progress in reading with teacher guidance. The student demonstrates 90–94% comprehension of text.

irony: The use of words to convey the opposite of their literal meaning: the words say one thing, but mean another. Often meant to reflect the author's tone or the attitude of a character or situation.

key word searches: A key term or phrase the computer uses in order to begin an online search for specific information.

language registry: The systematic differences of language use determined by regional, social or situational changes (e.g., a child might say "yup" at home, but would be expected to say "yes" at school).

letter patterns: Common letter groupings that represent specific sounds (e.g., *–ing* in "string" and *–ough* in "enough").

literary devices: Techniques used to convey or enhance an author's message or voice (e.g., idiom, figurative language, metaphor, exaggeration, dialogue, and imagery).

literary/narrative genres: Subcategories used to classify literary works, usually by form, technique, or content (e.g., novel, essay, short story, comedy, epic).

literary/narrative text: Text that describes action or events; usually includes a problem and resolution; usually, but not always, fiction.

Glossary

main idea: The gist of a passage; central thought; the chief topic of a passage which can be expressed or implied in a word or phrase; the topic sentence of a paragraph; a statement in sentence form which gives the stated or implied major topic of a passage and the specific way in which the passage is limited in content or reference.

mental imagery: Words or phrases that appeal to one or more of the five senses, allowing the reader to form mental pictures or images while reading.

metaphor: A figure of speech that compares two things without using the words *like* or *as* (e.g., laughter is the best medicine).

mood: The emotional state of mind expressed by an author or artist in his/her work, or the emotional atmosphere produced by an artistic work.

multiple-meaning words: Words with the same spelling and/or pronunciation which have more than one meaning depending on their context, such as "The wind blew" and "Please wind the clock."

non-technical documents: In this context, non-technical refers to documents (e.g., memos, lists, job applications) in which the content and vocabulary are not tied to a specific subject.

oddity tasks: In phonemic awareness, identifying which word in a set of three or four has the "odd" sound (e.g., run, rug, and toy).

onomatopoeia: A term used to describe words whose pronunciations suggest their meaning (e.g., meow, buzz). Words used to represent a sound.

onset and rime: Parts of spoken language that are syllables. An onset is the initial consonant(s) sound of a syllable (the onset of bag is *b*–; of swim, *sw*–). A rime is the part of the syllable that contains the vowel and all that follows it (the rime of bag is *–ag*; of swim, *–im*). Not all syllables or words have an onset, but they all have a rime (e.g., the word or syllable "out" is a rime without an onset).

oral language structure: Spoken language has five linguistic systems. They include the phonological (sounds of language), the syntactic (order and grammar), the semantic (meanings), the pragmatic (social interactive), and lexicon (vocabulary).

organizational features: Tools the author uses to organize ideas (e.g., caption and headings).

organizational structures: The organization of a text.

personification: A figure of speech in which nonhuman objects, such as ideas, objects, or animals, are given human characteristics (e.g., "flowers danced about the lawn").

persuasive devices: A technique the author uses to move the reader to his/her point of view, such as bias, overgeneralization, and association.

phoneme: The smallest unit of sound in a spoken word that makes a difference in the word's meaning.

phonemic awareness: The ability to hear, identify, and manipulate individual sounds (phonemes) in spoken words.

phonics: The understanding that there is a predictable relationship between phonemes (the sounds of spoken language) and graphemes (the letters and spellings that represent those sounds in written language).

phonological awareness: A general understanding of the sound structure of words, including rhymes, syllables, and phonemes.

plot: The structure of the events in a story, usually including rising action, climax, and resolution.

point of view: The perspective from which a story is told. The three points of view are first person, third person, and omniscient.

predict: To foresee what might happen in a text based on textual clues and a reader's background knowledge or schema.

predictions: Foretelling what might happen next in a story or poem by using textual clues and a reader's background knowledge or schema.

Glossary

prefix: An affix attached before a base word or root, such as re- in reprint. A prefix slightly alters the meaning of a root word. For example, reprint means to print again.

primary sources: The original source of resource information (e.g., letter, encyclopedia, book).

print conventions: The rules that govern the customary use of print in reading and writing including directionality of print, punctuation, and capitalization.

prior knowledge: The knowledge that stems from previous experience. Note: prior knowledge is a key component of the schema theory of reading comprehension.

propaganda: Written or oral presentations intended to persuade the audience to a particular point of view often by misrepresenting data or exaggerating the facts.

propaganda techniques: Methods used in creating propaganda, such as bandwagon, peer pressure, repetition, and testimonials/endorsements.

pull-down menu: A computer term that refers to a list of words that appears when the cursor is on a menu item. Also called a drop down list box.

questioning strategies: In these strategies, a reader asks questions about a text before, during, and after reading and then searches for answers (e.g., Question Answer Response (QAR); Survey, Question, Read, Recite, Review (SQ3R)).

root words: Meaningful base form of a complex word, after all affixes are removed. A root may be independent, or free, as "read" in unreadable, or may be dependent, or bound, as –liter– (from the Greek word for letter) in illiterate.

sarcasm: A remark used to "make fun of" or "put down" someone or something. The remark is not sincere and is often intended to hurt someone's feelings.

scan: To examine or read something quickly, but selectively, for a purpose.

scanning: Examining or reading something quickly, but selectively, for a purpose.

schema: The accumulated knowledge drawn from life experiences that a person has to help understand concepts, roles, emotions, and events.

secondary sources: Sources of information that are derived from primary or original sources (e.g., gossip).

segment: The act of separating the sounds in a word in order to assist decoding or spelling.

semantic mapping: A graphic display of a cluster of words that are meaningfully related.

sentence structure: Any of a number of basic sentence types in a language. The pattern or structure of word order in sentences, clauses, or phrases.

sequence: The arrangement or ordering of information, content, or ideas (e.g., chronological, easy to difficult, part to whole).

sequential: Marked by an arrangement or order of information, content, or ideas, such as part to whole, easy to difficult, etc.

setting: The time(s) and place(s) in which a narrative takes place.

short vowel sounds: The sound of /a/ as in cat, /e/ as in hen, /i/ as in fit, /o/ as in hot, and /u/ as in pup.

sight words: Words that are immediately recognized as wholes and do not require word analysis for identification.

similes: Figures of speech comparing two unlike things usually using *like* or *as* (e.g., Like ancient trees, we die from the top).

skim: To read or glance through quickly.

story elements: The critical parts of a story, including character, setting, plot, problem, solution. At upper grades, the terms problem and solution change to conflict and resolution.

Glossary

story structure: The pattern of organization in narration that characterizes a particular type of story.

structural analysis: The identification of word-meaning elements, such as "re–" and "read" in reread, to help understand the meaning of a word as a whole.

sub-genres: Genres within other genres (e.g., haiku is a sub-genre of poetry, and mystery is a sub-genre of fiction).

subplot: A minor collection of events in a novel or drama that have some connection with the main plot and should (1) comment on, (2) complicate/defeat, or (3) support the main plot.

suffix: An affix attached to the end of a base, root, or stem that changes the meaning or grammatical function of the word (e.g., –en added to ox to form oxen).

summarize: To determine what is important in the text, condense this information, and put it into the students' own words.

summary: A synthesis of the important ideas in a text presented in a condensed form.

syllabication: Division of words into syllables. A syllable is a word part that contains a vowel, or in spoken language a vowel sound (e-vent; news-pa-per; ver-y).

synonym: A word having a meaning similar to that of another word.

task-oriented text: Text written specifically to direct the reader as to how to complete a task.

technical: Content or vocabulary directly related to specific knowledge or information in a career or interest area.

text complexity: Text demands on the reader increase substantially throughout the grades. Items that influence complexity of text include: highly specialized vocabulary and concepts; abstract concepts presented with minimal context; increased concept load/density; readability considerations; and unique writing patterns in informational text.

text features: A prominent characteristic of a particular type of text, such as chapter titles, sub-headings, and bold-faced words in a history text.

text organizational structures: Expository text is structured in certain ways. The five text structures that students are most likely to encounter are cause-effect, compare/contrast, description, problem/solution, and chronological or time order.

theme: A topic; a major idea or proposition broad enough to cover the entire scope of a literary work. Note: a theme may be stated or implicit, but clues to it may be found in the ideas that are given special prominence or tend to recur in a work.

unfamiliar text: Unseen, unpracticed reading material.

vocabulary strategies: A systematic plan to increase understanding of words (e.g., categorizing and classifying, semantic mapping, semantic feature analysis, concept of definition maps, analogies, using the dictionary and other reference materials, using word parts, using morphemic analysis, using context clues).

word families: A collection of words that share common orthographic rimes (e.g., thank, prank, dank).

word recognition strategies: Strategies for determining the pronunciation and meaning of words in print.

Reading Tutorial

The Reading Practice Tutorial is made up of multiple-choice, short-answer, and extended-response questions. These questions show you how the skills you have learned in Reading class may be tested on the Reading MSP. The questions also give you a chance to practice your skills. If you have trouble with a question, talk with a parent or teacher.

Read each question carefully. If you do not know an answer, you may skip the question and come back to it later.

When you finish, check your answers.

Directions for the Reading Tutorial

Today you will take the Reading Tutorial. You will read passages and answer questions. You may look back at the passage when you are answering the questions.

Directions to the Student

There are three different types of questions on this assessment:

1. There are multiple-choice questions that require you to choose the best answer.
2. There are short-answer questions for which you will write phrases or sentences on the lines provided in your booklet.
3. There are some extended-response questions for which you are expected to write a longer and more detailed answer in your booklet.

Here are some important things to remember as you take this assessment:

1. Read each passage. You may look back at the reading passage as often as you want.
2. The paragraphs are numbered for all reading passages. A question about a particular paragraph will refer to the paragraph number.
3. Read each question carefully. When you write your answers, write them neatly and clearly on the lines provided. You may use sentences, phrases, paragraphs, lists, or charts to explain your ideas. Cross out or erase any part of your work you do not want to include as part of your answer.
4. When you choose a multiple-choice answer, make sure you completely fill in the circle next to the answer. Erase completely any marks that you want to change on multiple-choice items.
5. Use only a **No. 2 pencil**, not a mechanical pencil or pen, to write your answers. If you do not have a No. 2 pencil, ask your teacher to give you one.
6. You should have plenty of time to finish every question on the assessment. If you do not know the answer to a question, go on to the next question. You can come back to that question later.
7. When you reach the word **STOP** in your booklet, you have reached the end of the Reading Tutorial. Do **not** go on until you are told to turn the page.
8. If you finish early, you may check your work in this section **only**.

Go on ▶

Sample Questions

To help you understand how to answer the test questions, look at the sample test questions below. These questions do not refer to the selections you are about to read. They are included to show you what the questions in the test are like and how to mark or write your answers.

Multiple-Choice Sample Question

For this type of question you will select the answer and fill in the circle next to it.

1 What problem does Jake face?

- ○ A. The noise made by the insects keeps him up at night.
- ● B. He is scared of thunderstorms.
- ○ C. He doesn't like living on a ship.
- ○ D. The rain is so loud, he can not sleep.

For this sample question, the correct answer is B; therefore, the circle next to B is filled in.

Short-Answer Sample Question

For this type of question you will write a short answer consisting of a few phrases or sentences.

2 Based on the information in the story, predict what will most likely happen if there is another thunderstorm. Provide information from the story to support your prediction.

If there is another thunderstorm I think Jake will lie still in his warm bed and think about how the crickets are getting needed water from the storm. The story tells us that Jake stays in his bed during the storms. Then, at the end of the story Jake realizes that the water from the storms helps the crickets and this makes him feel better.

Go on ▶

Extended-Response Sample Question

For this type of question you will write a more extensive answer, offering more examples and more detail.

3 What four pieces of advice would Jake give about thunderstorms? Provide information from the story in your answer.

Jake would tell his friends it is okay to be frightened of thunderstorms. After all, they are very loud storms and can keep you up at night. Jake might suggest to his friends to stay in their bed where they are warm and safe during a storm. This is what Jake does when there is a storm at night. Jake would tell his friends that there are good things about thunderstorms, such as providing water for the crickets that live outside. Jake might suggest to his friends to think about these good things during the storm to help feel more relaxed.

Go on ▶

Directions: Read the speech and answer the questions.

Meeting of the Future Storm Chasers Club

1   Good afternoon, fellow students and future storm chasers. As this club's founder and president, I would like to welcome you to the first meeting of the Future Storm Chasers Club. I hope we can convince many more storm chasers to join us in the future!

2   Although we are too young to be storm chasers, it is not too soon to begin preparing for this thrilling hobby. In the meetings of this club, we will talk about the different things storm chasers do. We have lined up some storm chasers as guest speakers. They will not only describe what they do as storm chasers, they will show pictures and video they have taken while out chasing storms. The images are really neat. I've met with some of these people, and they know a lot about storm chasing.

3   Another part of storm chasing we will talk about is safety. Storm chasing is an extremely dangerous activity, and it is very important to understand how to stay safe. We want to live to tell others what we have seen. Storm chasers often get much closer to tornadoes and other forms of severe weather than most people would choose to, so they have to know how to handle dangerous situations. Some of our speakers will be covering this issue in their presentations, so you should pay close attention.

4   Our speaker next week will be Jim Jeffries, a storm chaser who has followed tornadoes all over the Midwest. Jeffries has been chasing storms for over 10 years. He likes the activity so much, he's gotten his wife and brother involved in storm chasing. Together, they have chased hundreds of storms. He knows so much about this topic, you'll be amazed at the information he will be able to share. Plus, he has great safety tips. Anyone who would like to hear his fascinating speech should be here at 3:00 p.m. next Wednesday. Tell your friends and family. Everyone is welcome.

5   Thank you all for coming to this meeting. I hope to see everyone next week. Don't forget to spread the word. We want all future storm chasers to join the club!

Go on ▶

1 Which word could the author have used in paragraph 3 instead of *severe*?

- A. Tame
- B. Unrelenting
- C. Bad
- ● **D. Heavy**

2 What is the main idea of the speech?

- A. A student is trying to increase the membership of the Future Storm Chasers Club.
- ● **B. A student is giving information about the guest speaker for the Future Storm Chasers Club meeting.**
- C. A student is explaining why the Future Storm Chasers Club has been founded and its purpose.
- D. A student is explaining the dangers of storm chasing to members of the Future Storm Chasers Club.

Go on ▶

3 Based on the information in the speech, what do you predict the leader of the Future Storm Chasers Club will do now that the first meeting has concluded?

- ○ A. The club leader will disband the club because of a lack of interest.
- ○ B. The club leader will resign because the position requires too much effort.
- ○ C. The club leader will ask the science teacher to take over the club duties.
- ● D. The club leader will promote next week's guest speaker and encourage people to attend the meeting.

4 Write a summary of the speech. Include **three** main events from the speech in your summary.

In the speech the speaker tells about what the members will be doing in the club. Next, he talked about the dangers of being a Storm Chaser. Finally, he told the group to meet in the same place at three P.M. next Wendesday to hear Jim Jeffries speach to learn more about Storm Chasers.

Go on ▶

Directions: Read the story and answer the questions.

The Exchange Student

1   "When does her plane arrive?" Emily asked her mother impatiently. Emily wasn't very good at hiding her excitement. She knew her mother did not want to hear her say one more word, but Emily didn't care. She was as bouncy as a gymnast on a trampoline.

2   "For the last time, we'll leave for the airport around two o'clock. Misa's flight doesn't get in until three this afternoon," replied Mrs. Norris, who was obviously annoyed by her daughter's repeated questioning. "I know you're excited, but this is Misa's first time in the United States. We want to make her feel right at home here in New York," Emily's mother reminded her.

3   The Norris family had agreed to host a foreign exchange student for the summer. The International Student Exchange, a program sponsored by Emily's school, had invited students from around the world to enjoy two months in New York. Emily had been waiting for the exchange student to arrive ever since. Today, the waiting would be over. Misa was traveling from Italy on Flight 223 out of Milan to America.

4   Mrs. Norris had prepared the extra bedroom for Misa. It was a few steps down the hall from Emily's room. Emily gave the room one last glance as two o'clock approached. She wanted to make sure everything was in order for Misa's arrival.

Go on ▶

5   Emily and her mother reached the airport. They found a flight schedule posted on an electronic display board. As Mrs. Norris searched for Misa's flight, a voice on the airport speaker announced that Flight 223 from Italy had landed. Emily's stomach filled with butterflies. The moment she had been waiting for had finally arrived. Several transatlantic passengers, looking tired and anxious, walked past. Emily looked around each person, trying to spot Misa.

AIRLINE FLIGHT NO.	DEPARTURE CITY	ARRIVAL TIME	STATUS
King Jet 687	Berlin	2:30 p.m.	Delayed
International 456	London	2:45 p.m.	On time
Continuous 223	Milan	3:00 p.m.	On time
Fly Away 313	Houston	4:05 p.m.	Delayed
Air Ways 725	Washington	5:25 p.m.	On time

6   "I think I see her! I think I see her!" Emily said as she tugged on her mom's arm.

7   "Hello, Misa!" Emily waved. "Welcome to the United States!" The girls gave each other hugs. Misa asked question after question, while Emily answered question after question. Mrs. Norris smiled. *At least Misa has as many questions as Emily*, she thought to herself.

Go on ▶

5 What is the purpose of the flight chart in the selection?

- ○ **A.** To compare Misa's flight with other arriving flights
- ○ **B.** To explain that only international flights are on time
- ○ **C.** To show that Misa's flight is on time
- ○ **D.** To reveal all of the other delayed airport flights

6 Based on the story, which word best describes Emily?

- ○ **A.** Enthusiastic
- ○ **B.** Timid
- ○ **C.** Scared
- ○ **D.** Worried

Go on▶

7 What are **two** differences between Emily and Misa? Include information from the story in your answer.

_____

_____

_____

_____

_____

_____

_____

_____

Go on ▶

8 Which **two** objects does the author compare using a simile?

- A. Gymnast and trampoline
- B. Emily and a gymnast
- C. Emily and a trampoline
- D. Misa and a gymnast

9 Which statement is the most important conclusion the reader can draw from the story?

- A. Emily is not looking forward to having Misa in her house for two months.
- B. Emily's mother is curious to see why Misa left Italy.
- C. Misa is scared about being in the United States without her family.
- D. Emily and Misa will enjoy learning about each other over the next two months.

Go on▶

Directions: Read the selection and answer the questions.

A Dangerous Hobby

1   Many people fear "Mother Nature" and the unpredictable weather she has been known to cause, but there are those who are fascinated by her. They want to see her effects up close. As a matter of fact, they go out looking for bad weather and storms. These people are known as storm chasers.

2   Storm chasers are everyday people who are awestruck by severe storms and extreme weather. They are so fanatical that they actually go out in search of it! They sometimes travel hundreds of miles from state to state in search of the big storm. Certain areas of the United States, including the Great Plains and the Midwest, are known to produce such storms. As a result, many storm chasers come from those areas or settle there.

3   Usually, the main goal of the storm chaser is to witness a tornado, but this isn't always the case. Oftentimes, the severe weather becomes dangerous. The storm chasers could be risking their own safety if they continue to chase the storm. When the risk is too great, storm chasers have to give up the chase. For most, however, the real reward is the total experience of the storm itself and getting to see nature's awesome power up close.

4   Technological advances have given meteorologists, scientists who study weather, the ability to predict certain weather patterns and storms. Meteorologists are people who have earned a special degree from a university and who work to forecast and report the weather.

5   Today, storm chasers have many more tools and information to work with than the first chasers did. Using this information, storm chasers are likely to experience more storms than ever before. While most storm chasers are there for the experience and to take photographs or to shoot video, some do collect meteorological information for weather research.

6   Storm chasers come from all walks of life, but their interest in weather and storms gives them something in common. As long as there is weather to observe, storm chasers will continue to follow the skies.

Go on ▶

10 What is the author's purpose for writing the selection? Provide **three** details from the selection to support your answer.

11 Which sentence from the selection is an opinion?

    ○ **A.** "These people are known as storm chasers."

    ○ **B.** "They are so fanatical that they actually go out in search of it!"

    ○ **C.** "They want to see her effects up close."

    ○ **D.** "They sometimes travel hundreds of miles from state to state in search of the big storm."

12 In paragraph 4 of the selection, why does the author include factual information about meteorologists?

    ○ **A.** To emphasize that storm chasers do not have college degrees

    ○ **B.** To emphasize that storm chasers can forecast the weather

    ○ **C.** To emphasize that storm chasers and meteorologists have the same interests

    ○ **D.** To emphasize that meteorologists are different than storm chasers

Go on ▶

13 Based on the information in the selection, which generalization can the reader make about chasing storms?

- ○ A. Storm chasing is an important part of daily life.
- ○ B. Storm chasing can give scientists insight into weather patterns.
- ○ C. Storm chasing can be a dangerous, unfulling activity.
- ○ D. Storm chasing is only done by people with college degrees.

Go on ▶

14 How do the author's feelings about storm chasing in "Meeting of the Future Storm Chasers Club" on page 35 compare to the author's feelings about storm chasing in "A Dangerous Hobby" on page 43? Include **one** detail from the speech and **one** detail from the selection in your answer.

Go on▶

15 Many people like to chase storms, regardless of the dangers. Would the author of "A Dangerous Hobby" think this is a good idea? Provide **two** details from the selection to support your answer.

Reading Assessment One

Introduction

Reading Assessment One is made up of multiple-choice, short-answer, and extended-response questions. These questions show you how the skills you have learned in Reading class may be tested on the Reading MSP. The questions also give you a chance to practice your skills. If you have trouble with a question, talk with a parent or teacher.

Read each question carefully. If you do not know an answer, you may skip the question and come back to it later.

When you finish, check your answers.

Directions for Reading Assessment One

Today you will take the Reading Assessment One. You will read passages and answer questions. You may look back at the passage when you are answering the questions.

Directions to the Student

There are three different types of questions on this assessment:

1. There are multiple-choice questions that require you to choose the best answer.
2. There are short-answer questions for which you will write phrases or sentences on the lines provided in your booklet.
3. There are some extended-response questions for which you are expected to write a longer and more detailed answer in your booklet.

Here are some important things to remember as you take this assessment:

1. Read each passage. You may look back at the reading passage as often as you want.
2. The paragraphs are numbered for all reading passages. A question about a particular paragraph will refer to the paragraph number.
3. Read each question carefully. When you write your answers, write them neatly and clearly on the lines provided. You may use sentences, phrases, paragraphs, lists, or charts to explain your ideas. Cross out or erase any part of your work you do not want to include as part of your answer.
4. When you choose a multiple-choice answer, make sure you completely fill in the circle next to the answer. Erase completely any marks that you want to change on multiple-choice items.
5. Use only a **No. 2 pencil**, not a mechanical pencil or pen, to write your answers. If you do not have a No. 2 pencil, ask your teacher to give you one.
6. You should have plenty of time to finish every question on the assessment. If you do not know the answer to a question, go on to the next question. You can come back to that question later.
7. When you reach the word **STOP** in your booklet, you have reached the end of the Reading Assessment One. Do **not** go on until you are told to turn the page.
8. If you finish early, you may check your work in this section **only**.

Go on▶

Directions: Read the story and answer the questions.

The New Scooter

1   A.J. tugged at the wrapping paper. Each tear revealed more of what he had hoped to see. Slowly, a brand new scooter made its way out of the box. A fresh-cut piece of chocolate cake sat untouched on the table. All A.J. could think about was trying out his new scooter. The raindrops tapping against the windowpane told him his first adventure would have to wait for another day.

2   The scooter may have been a birthday gift, but to A.J. it was much more than that. A.J. had asked for the same scooter for his last birthday; he had not gotten it. Instead, he had gotten some clothes and a speech from his parents about how important it was to do well in school. For the next year, A.J. worked as hard as he could in every class. He made the honor roll every quarter. A.J. looked at the scooter and smiled. To him, it stood for all the homework he had worked so hard on.

3   Another stormy day passed before A.J. saw light break through the dark clouds. The shiny scooter rested against the sun-porch railing. The fluorescent green wheels popped with color. The metallic shaft and handlebars glimmered in the light. A.J. grabbed his new toy and pushed both himself and the scooter outside with excitement. With one foot on the scooter's base, he used his other leg to give the scooter a gentle push. He then lifted his other leg to the scooter base and drifted down his driveway. There were several other kids outside enjoying the sun. It wasn't long before A.J. saw an old friend.

4   "Hey, Ann!" A.J. cried out. "What do you think of my new scooter?" He hardly gave her enough time to answer. He spun the scooter around on its front wheel, and he was off in another direction.

5   When Ann was able to get close to the duo, she admired A.J.'s new set of wheels. She had seen a scooter just like it at the bike shop, but the one she wanted had purple wheels. Ann watched as A.J. raced up and down the concrete driveway. It was as if he had been riding all his life.

6   "You're pretty good at making that thing move," she said as he whizzed past her for the second time. A.J. stopped in front of Ann and decided this was the perfect opportunity to impress her with some tricks.

Go on ▶

7   "Watch this!" A.J. said while he rode the scooter with just one hand. The more he rode, the braver he became. Holding onto the handlebars tightly, he made the scooter hop. Next, he lifted his right leg forward; then, he swung it behind him. He steered the scooter to the right and to the left, all while balancing on one leg. A.J. hardly let the scooter lose momentum before he would propel it forward again.

8   "Please be careful! You should be wearing a helmet if you're going to do tricks like that," Ann reminded him, but he didn't seem to pay attention to her.

9   It wasn't until A.J. felt like increasing the difficulty of his stunts that he actually let go of the scooter. He dropped it in the grass when he found some wooden boards lying by the side of the garage. He arranged them carefully until they formed a small ramp like the one he had seen in the bicycle magazines he kept in his room. "I'm ready for some action now," A.J. cried out, hoping Ann was watching him.

10   The first time over the ramp seemed easy, and A.J.'s confidence grew. "That was great!" Ann said, smiling.

11   She didn't want to admit it, but she hoped he would ask her if she wanted to ride his scooter. She wanted to try a trick or two, but A.J. wouldn't let go of his new scooter.

12   He wheeled the scooter around for his second attempt. "This one will be even better," he shouted.

13   A.J. zipped up and down the driveway and made his way toward the ramp. Just as the front wheel reached the board, the scooter flipped, leaving A.J. on the hard pavement. Ann rushed over and saw that A.J. had some scrapes and a large knot on his forehead. "Are you OK?" she said.

14   A.J. sat up slowly, rubbing his head. "On a scale of one to ten, what would you give me for that stunt?" asked A.J. It was his way of saying, "I'm fine." They both started laughing.

15   "I'll give you a ten if you promise not to do that again until you've got a helmet."

16   "It's a deal," said A.J. He climbed to his feet and looked around for the misplaced scooter. Upon careful examination, nothing seemed to be out of place. Although he knew he needed to take a break from the scooter, A.J. couldn't wait to try it again!

Go on ▶

1 Based on the story, which word best describes A.J.?

- ○ **A.** Boring
- ○ **B.** Scared
- ○ **C.** Lonely
- ○ **D.** Adventurous

2 Any of these titles could be another title for the story. Choose the title you think best fits the story.

His New Scooter

The Scooter Adventure

The Perfect Reward

Provide **two** details from the story to support your choice.

_____

_____

_____

_____

_____

_____

_____

Go on ▶

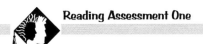

3 What is the meaning of the word *opportunity* as it is used in paragraph 6 of the story?

○ A. To take a chance

○ B. To be lucky

○ C. A fit time

○ D. An unfit time

4 In paragraph 8 of the story, why does the author include foreshadowing?

○ A. To let the reader know that Ann was not concerned about A.J.'s safety

○ B. To let the reader know that Ann was worried about A.J.'s safety

○ C. To let the reader know that A.J. was not concerned about his safety

○ D. To let the reader know that A.J. was worried about his safety

Go on ▶

5 Which statement is the most important conclusion the reader can draw from the selection?

- ○ **A.** A.J. will continue to work hard to reach all of his goals.
- ○ **B.** A.J. assumes that every thing will be given to him.
- ○ **C.** A.J. will continue to ask for a different gift each year for his birthday.
- ○ **D.** A.J. will reach his goals no matter what he does.

6 What is the meaning of the word *fluorescent* as it is used in paragraph 3 of the story?

- ○ **A.** Dull
- ○ **B.** Bright
- ○ **C.** Shiny
- ○ **D.** Cloudy

7 Why does A.J. promise to wear a helmet at the end of the story?

- ○ **A.** The scooter flips over, leaving A.J. on the pavement.
- ○ **B.** Ann wants to try to do tricks on the scooter.
- ○ **C.** A.J. rides his scooter over the ramp and down the street.
- ○ **D.** Ann told A.J. to put on a helmet before trying difficult stunts.

Go on ▶

Directions: Read the selection and answer the questions.

Duke Ellington
(1899–1974)

1 Duke Ellington was born Edward Kennedy Ellington on April 29, 1899, in Washington, D.C. His mother and father, Daisy Kennedy Ellington and James Edward Ellington, were kind and loving parents. They enjoyed music and shared their love of the art with their son. During Ellington's youth, many who knew him admired his noble qualities. As a result, they gave him a regal-sounding nickname: "The Duke." Person after person called Ellington "The Duke" and the nickname stuck. Future generations have come to know "The Duke" as one of history's greatest jazz talents.

2 By the age of seven, Ellington was taking piano lessons, but he was not impressed with the instrument, not even a little bit. Instead, Ellington was fascinated by the game of baseball. In fact, for his very first job, he sold packages of peanuts at Washington Senators' baseball games. Although he didn't realize it at the time, this experience made a big impact on Ellington. As a shy kid, he was terrified at the thought of facing large crowds. Selling peanuts at the stadium helped Ellington overcome his fears. He had to get comfortable with large numbers of people in order to sell his peanuts and make money. As time progressed, he grew to like "performing" for the crowds.

3 Even though he didn't like the piano at first, Ellington was slowly drawn to music. To study the arts, he attended the Armstrong Manual Training School, rather than attending a typical high school. During a family trip to Asbury Park in New Jersey, Ellington came to know Harvey Brooks, a dynamic piano player. Brooks was quick to show Ellington his signature piano tricks and shortcuts. Ellington felt Brooks' influence immediately, as Brooks made quite an impression on the young Ellington. Ellington later recalled his experiences with Brooks, "When I got home I had a real yearning to play. I hadn't been able to get off the ground before, but after hearing him I said to myself, 'Man you're going to have to do it.'" From then on, he spent most of his free time listening to ragtime pianists. He was amazed at the way they could charm audiences. He was drawn to their music and charm. Although inspired by their performances, Ellington was able to pound out his own unique style.

Go on ▶

4   Ellington was composing his own works by the age of 17. Throughout his late teens and early twenties, he enjoyed modest success as a professional musician. He was earning enough to make a living, and he was making his way doing something he had grown to love. In 1923, Ellington left Washington, D.C., and headed for New York City. Ellington and his band, The Washingtonians, began recording his music and playing in popular nightclubs. The band was featured regularly on a national radio broadcast titled "From the Cotton Club." The power and beauty of Ellington's work was recognized all over the nation, and his popularity increased.

5   Eventually, Ellington took his band on the road. They played all over the world, from London to Egypt to Los Angeles. Their music was well received as it poured from concert halls, nightclubs, and theaters. Ellington was not only a musician, a bandleader, and a songwriter, he was also a global ambassador for American music.

6   Ellington's talents were honored many times over. The French government presented him with their highest award, the Legion of Honor. The government of the United States granted him its highest civil honor, the Presidential Medal of Freedom. Ellington received honorary degrees from Howard University and Yale University. He was given membership in the American Institute of Arts and Letters, and he was the first jazz musician to become a member of the Royal Music Academy in Stockholm.

7   While Ellington is largely recognized for his contributions to big-band jazz, he had many other achievements. He composed over 1,500 works, and many were featured in musicals, films, television programs, concerts, and ballets. His talents as a composer and as a dynamic performer have kept Ellington's legend alive for almost a century. Duke Ellington died from cancer on May 24, 1974. He is considered one of the greatest musicians of all time. As a tribute to his impact on music, his musical materials are now preserved in the Duke Ellington Collection at the Smithsonian Institute in Washington, D.C.

Go on ▶

8   What is most likely the author's purpose for writing this selection?

○ A.   To inform the reader about an important American jazz musician

○ B.   To explain the history of American jazz to the reader

○ C.   To demonstrate the story of Ellington's career as a global ambassador of jazz music to the reader

○ D.   To entertain the reader with the story of how Duke Ellington received his nickname

9   What is the meaning of the word *impact* as it is used in paragraph 7 of the selection?

○ A.   The force of a collision

○ B.   An impression left by an event

○ C.   Violent contact

○ D.   A shock

Go on ▶

10 What does the author mean when he/she says, "Ellington was slowly drawn to music" in paragraph 3 of the selection?

○ A. Ellington was an artist as well as a musician.

○ B. Ellington was tied in a contest with another musician.

○ C. Ellington was slowly attracted to playing music.

○ D. Ellington looked tired after playing music.

11 Which sentence best explains why Duke Ellington was a talented composer?

○ A. "He is considered one of the greatest musicians of all time."

○ B. "He composed over 1,500 works, and many were featured in musicals, films, television programs, concerts, and ballets."

○ C. "Future generations have come to know 'The Duke' as one of history's greatest jazz talents."

○ D. "Ellington was not only a musician, a bandleader, and a songwriter, he was also a global ambassador for American music."

Go on▶

12 Write a summary of the selection. Include **three** important ideas from the selection in your summary.

_____

_____

_____

_____

_____

_____

_____

_____

_____

13 According to the selection, what happened when Ellington saw piano performances by Harvey Brooks?

- **A.** Ellington decided to become a member of the Washingtonians.
- **B.** Ellington started to attend the Armstrong Manual Training School.
- **C.** Ellington was influenced by his performance style, and his interest in piano playing had been sparked.
- **D.** Ellington started taking piano lessons from Brooks, who helped Ellington create his own style.

Go on ▶

Directions: Read the story and answer the questions.

from **Little Women**
by Louisa May Alcott

1   "Christmas won't be Christmas without any presents," grumbled Jo, lying on the rug.

2   "It's so dreadful to be poor!" sighed Meg, looking down at her old dress.

3   "I don't think it's fair for some girls to have plenty of pretty things, and other girls nothing at all," added little Amy, with an injured sniff.

4   "We've got Father and Mother, and each other," said Beth contentedly from her corner.

5   The four young faces on which the firelight shone brightened at the cheerful words, but darkened again as Jo said sadly, "We haven't got Father, and shall not have him for a long time." She didn't say "perhaps never," but each silently added it, thinking of Father far away, where the fighting was.

6   Nobody spoke for a minute; then Meg said in an altered tone, "You know the reason Mother proposed not having any presents this Christmas was because it is going to be a hard winter for everyone; and she thinks we ought not to spend money for pleasure, when our men are suffering so in the army. We can't do much, but we can make our little sacrifices, and ought to do it gladly. But I am afraid I don't," and Meg shook her head, as she thought regretfully of all the pretty things she wanted.

7   "But I don't think the little we should spend would do any good. We've each got a dollar, and the army wouldn't be much helped by our giving that. I agree not to expect anything from Mother or you, but I do want to buy *Undine and Sintran* for myself. I've wanted it so long," said Jo, who was a bookworm.

8   "I planned to spend mine in new music," said Beth, with a little sigh, which no one heard but the hearth brush and kettle-holder.

9   "I shall get a nice box of Faber's drawing pencils; I really need them," said Amy decidedly.

10   "Mother didn't say anything about our money, and she won't wish us to give up everything. Let's each buy what we want, and have a little fun; I'm sure we work hard enough to earn it," cried Jo, examining the heels of her shoes in a gentlemanly manner.

Go on ▶

11 "I know I do—teaching those tiresome children nearly all day, when I'm longing to enjoy myself at home," began Meg, in the complaining tone again.

12 "You don't have half such a hard time as I do," said Jo. "How would you like to be shut up for hours with a nervous, fussy old lady, who keeps you trotting, is never satisfied, and worries you till you're ready to fly out the window or cry?"

13 "It's naughty to fret, but I do think washing dishes and keeping things tidy is the worst work in the world. It makes me cross, and my hands get so stiff, I can't practice well at all." And Beth looked at her rough hands with a sigh that any one could hear that time.

14 "I don't believe any of you suffer as I do," cried Amy, "for you don't have to go to school with impertinent girls, who plague you if you don't know your lessons, and laugh at your dresses, and label your father if he isn't rich, and insult you when your nose isn't nice."

15 "If you mean libel, I'd say so, and not talk about labels, as if Papa was a pickle bottle," advised Jo, laughing.

16 "I know what I mean, and you needn't be statirical about it. It's proper to use good words, and improve your vocabilary," returned Amy, with dignity.

17 "Don't peck at one another, children. Don't you wish we had the money Papa lost when we were little, Jo? Dear me! How happy and good we'd be, if we had no worries!" said Meg, who could remember better times.

18 "You said the other day you thought we were a deal happier than the King children, for they were fighting and fretting all the time, in spite of their money."

19 "So I did, Beth. Well, I think we are. For though we do have to work, we make fun of ourselves, and are a pretty jolly set, as Jo would say."

Go on ▶

14 Based on the information in the story, what inference can the reader make about Amy's favorite hobby?

- ○ A. Amy likes to read.
- ○ B. Amy likes to play piano.
- ○ C. Amy likes to sew at home.
- ○ D. Amy likes to create artwork.

15 What is the main conflict in the story?

- ○ A. The sisters are trying to decide what they are getting for Christmas.
- ○ B. The sisters are trying to decide what they should buy for Christmas.
- ○ C. The sisters are trying to decide how to spend the Christmas money they received as gifts.
- ○ D. The sisters are trying to decide how much money to give to the army for Christmas.

Go on ▶

16 Is the sisters' decision to buy gifts for themselves a good idea? Provide **two** details from the story to support your answer.

_____

_____

_____

_____

_____

_____

_____

Go on ▶

17 What **four** pieces of advice would the sisters give about the best way to spend money? Provide information from the story in your answer.

18 What is the meaning of the word *impertinent* as it is used in paragraph 14 of the story?

- ○ **A.** Rude
- ○ **B.** Unimportant
- ○ **C.** Angry
- ○ **D.** Friendly

19 Write a summary of the story. Include **three** main events from the story in your summary.

_____

_____

_____

_____

_____

_____

_____

_____

_____

Go on ▶

20 Which sentence explains why the girls are worrying about Christmas gifts in the story?

- A. " 'We've got Father and Mother, and each other,' said Beth contentedly from her corner."

- B. "You know the reason Mother proposed not having any presents this Christmas was because it is going to be a hard winter for everyone; and she thinks we ought not to spend money for pleasure, when our men are suffering so in the army."

- C. "I agree not to expect anything from Mother or you, but I do want to buy *Undine and Sintran* for myself."

- D. "We can't do much, but we can make our little sacrifices, and ought to do it gladly."

Go on ▶

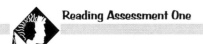

Directions: Read the poem and answer the questions.

Birches
by Robert Frost

When I see birches bend to left and right
Across the lines of straighter darker trees,
I like to think some boy's been swinging them.
But swinging doesn't bend them down to stay.
5   Ice-storms do that. Often you must have seen them
Loaded with ice a sunny winter morning
After a rain. They click upon themselves
As the breeze rises, and turn many-colored
As the stir cracks and crazes their enamel.
10 Soon the sun's warmth makes them shed crystal shells
Shattering and avalanching on the snow-crust—
Such heaps of broken glass to sweep away
You'd think the inner dome of heaven had fallen.
They are dragged to the withered bracken[1] by the load,
15 And they seem not to break; though once they are bowed
So low for long, they never right themselves:
You may see their trunks arching in the woods
Years afterwards, trailing their leaves on the ground
Like girls on hands and knees that throw their hair
20 Before them over their heads to dry in the sun.
But I was going to say when Truth broke in
With all her matter-of-fact about the ice-storm
(Now am I free to be poetical[2]?)
I should prefer to have some boy bend them
25 As he went out and in to fetch the cows—
Some boy too far from town to learn baseball,
Whose only play was what he found himself,
Summer or winter, and could play alone.
One by one he subdued his father's trees
30 By riding them down over and over again
Until he took the stiffness out of them,
And not one but hung limp, not one was left
For him to conquer. He learned all there was
To learn about not launching out too soon

---

[1]*bracken: a large coarse fern*
[2]*poetical: skilled or fond of poetry*

Go on ▶

35 And so not carrying the tree away
   Clear to the ground. He always kept his poise
   To the top branches, climbing carefully
   With the same pains you use to fill a cup
   Up to the brim, and even above the brim.
40 Then he flung outward, feet first, with a swish,
   Kicking his way down through the air to the ground.
   So was I once myself a swinger of birches.
   And so I dream of going back to be.
   It's when I'm weary of considerations,
45 And life is too much like a pathless wood
   Where your face burns and tickles with the cobwebs
   Broken across it, and one eye is weeping
   From a twig's having lashed across it open.
   I'd like to get away from earth awhile
50 And then come back to it and begin over.
   May no fate willfully misunderstand me
   And half grant what I wish and snatch me away
   Not to return. Earth's the right place for love:
   I don't know where it's likely to go better.
55 I'd like to go by climbing a birch tree,
   And climb black branches up a snow-white trunk
   Toward heaven, till the tree could bear no more,
   But dipped its top and set me down again.
   That would be good both going and coming back.
60 One could do worse than be a swinger of birches.

Go on ▶

21 Which two objects does the poet compare using a simile?

　○ **A.** Birches and ice storms

　○ **B.** Birches and girls drying their hair in the sun

　○ **C.** Birches and boys who swing in trees

　○ **D.** Birches and boys who play baseball

22 Which sentence tells how the narrator and a boy who lived "too far from town to learn baseball" are similar?

　○ **A.** "One by one he subdued his father's trees/By riding them down over and over again/Until he took the stiffness out of them,/And not one but hung limp, not one was left/For him to conquer."

　○ **B.** "And so I dream of going back to be."

　○ **C.** "So was I once myself a swinger of birches."

　○ **D.** "I should prefer to have some boy bend them/As he went out and in to fetch the cows—/Some boy too far from town to learn baseball,/Whose only play was what he found himself,/Summer or winter, and could play alone."

Go on▶

23 Explain why birches bend to the left and right. Include **two** details from the poem in your answer.

_____

_____

_____

_____

_____

_____

_____

_____

24 In lines 7–9 of the poem, why does the poet include alliteration?

○ A. To make the reader hear the ice on the branches

○ B. To make the reader see the ice on the branches

○ C. To make the reader smell the ice on the branches

○ D. To make the reader feel the ice on the branches

Go on▶

25 Based on the poem, which word best describes the narrator?

- A. Elated
- B. Miserable
- C. Longing
- D. Jubilant

26 What is the author's purpose for writing the poem?

- A. To inform
- B. To entertain
- C. To persuade
- D. To describe

Go on ▶

Directions: Read the selection and answer the questions.

Chewing Gum

1   Did you know that the discovery of chewing gum might date back to 50 A.D. when ancient people chewed the resin from trunks of trees? Historians have found that the Greeks chewed tree resin from the mastic tree. They called this early chewing gum "mastiche." Back then, gum was thought to clean teeth and to freshen breath. The Greeks weren't the only ones to chew the resin substance, though. Throughout Central America, the Mayan people chewed sap from the sapodilla tree; they called this sap "chicle." Also, North American Indians were known to chew the sap from spruce trees.

2   It wasn't until 1848 that the first gum was manufactured and sold to the public. John Curtis and his brother produced spruce gum, which they called "State of Maine Pure Spruce Gum." Sales were slow. It wasn't until a more popular type of gum came to be that the Curtis brothers found success with their idea. In 1850, they added paraffin gum to their product line. The paraffin gum had a better texture than the spruce gum, and it could be produced in a variety of flavors. The improvements satisfied people's tastes, and the public began to buy more gum.

3   Several different inventors experimented with the flavor and consistency of chewing gum over time. The first person to obtain a patent for chewing gum was William Finley Semple of Mount Vernon, Ohio. His 1869 patent stated his product was a "combination of rubber with other articles, in any proportions adapted to the formation of an acceptable chewing gum." It is interesting to note that Semple never manufactured chewing gum for the public.

4   During the late 1860s, one-time president of Mexico Antonio Lopez de Santa Anna settled in New Jersey. Santa Anna brought a large amount of Mexican chicle to the United States. He hoped he would be able to sell it. He met Thomas Adams of Staten, New York, and discussed the idea of using the chicle to invent a rubber-like substance that could be used to make carriage tires. Adams had been working as a photographer, but he also dabbled in many other trades. The idea of the new invention interested Adams, so he agreed to buy Santa Anna's chicle supply.

Go on ▶

5 Adams experimented with the chicle, but each time he was disappointed with his results. In fact, he almost threw away the entire supply. By accident, an idea sparked in Adams: he would use the chicle to create a chewing gum. His development was successful, and Adams was soon selling chewing gum in many local stores. By 1871, Adams had developed and patented a gum-producing machine. Once sold in lumps or chunks, Adams' gum now sold in sticks. His first flavored gum was made with licorice flavoring. He called this type of gum "Black Jack." In some parts of the country, this brand can still be found. Another of Adams' chewing gums, Tutti-Frutti, became the first chewing gum to be sold in a New York City subway station vending machine.

6 While Adams was able to give his gums some flavor, their tastiness was often lost after just a few moments of chewing. Around 1880, John Colgan of Louisville, Kentucky, had an idea for how to extend gum's flavor. He added flavoring to the sugar before the sugar was added to the chicle. This way, the gum's flavoring lasted longer. Colgan's Taffy Tolu Chewing Gum was a quick success.

Go on ▶

27 What is the meaning of the word *consistency* as it is used in paragraph 3 of the selection?

○ **A.** Firmness

○ **B.** Reliable

○ **C.** Compatible

○ **D.** Temperature

28 Which sentence best explains why paraffin gum was an important step for the Curtis brothers?

○ **A.** "It wasn't until 1848 that the first gum was manufactured and sold to the public."

○ **B.** "John Curtis and his brother produced spruce gum, which they called 'State of Maine Pure Spruce Gum.'"

○ **C.** "The improvements satisfied people's tastes, and the public began to buy more gum."

○ **D.** "In 1850, they added paraffin gum to their product line."

Go on▶

29 The author's purpose for writing the selection "Chewing Gum" may have been to inform the reader. Provide **two** details from the selection to support this purpose.

Directions: Read the selection and answer the questions.

Bubble Gum

1   It wasn't until shortly after the turn of the twentieth century that bubble gum came to be. In 1906, Frank Fleer, of the Fleer Company, invented a form of bubble gum named "Blibber-Blubber." Unfortunately, this bubble gum was so sticky it never made it to store shelves. It wasn't until many years later that Walter Diemer, an accountant  for Fleer's chewing gum company, created a new bubble gum while experimenting with different recipes for fun. What he developed was gum that was not as sticky and that seemed to stretch more than the regular chewing gum. This stretching characteristic was perfect for blowing bubble gum bubbles. Diemer's gum was colored pink because it was the only color the Fleer Company had on hand at the time. Diemer thought the pink color made his bubble gum more appealing to the eye, and to this day, most bubble gum is pink.

2   Over the past century, the popularity of bubble gum has grown and grown. Stores now sell bubble gum in all shapes, sizes, and flavors. Gum comes in shredded pieces, in long strips rolled like tape, and in squeezable tubes. Popular flavors include watermelon, mint, cotton candy, and cherry, just to name a few. Next time you're in the mood for some gum, you may face a selection of more than 30 different varieties. Each time you chew on a piece of gum, just think, you're enjoying something that has developed and changed over the past 2,000 years. Who knows what the future of gum will bring!

Go on ▶

30 Which sentence best explains why Blibber-Blubber never became popular?

- A. "It wasn't until shortly after the turn of the century that bubble gum came to be."
- B. "Diemer's gum was colored pink because it was the only color the Fleer Company had on hand at the time."
- C. "In 1906, Frank Fleer, of the Fleer Company, invented a form of bubble gum named 'Blibber-Blubber.'"
- D. "Unfortunately, this bubble gum was so sticky it never made it to store shelves."

31 Which word could the author have used in paragraph 1 instead of *experimenting*?

- A. Carrying
- B. Researching
- C. Moving
- D. Studying

Go on▶

32 The author of "Chewing Gum" states that the improvements in chewing gum "satisfied people's tastes, and the public began to buy more gum." Provide **two** examples from "Bubble Gum" that supports this statement.

Go on ▶

33 According to the selection, what are **two** similarities between chewing gum and bubble gum? Provide information from the selection in your answer.

_____

_____

_____

_____

_____

_____

_____

_____

According to the selection, what are **two** differences between chewing gum and bubble gum? Provide information from the selection in your answer.

_____

_____

_____

_____

_____

_____

_____

_____

STOP

Reading Assessment Two

Introduction

Reading Assessment Two is made up of multiple-choice, short-answer, and extended-response questions. These questions show you how the skills you have learned in Reading class may be tested on the Reading MSP. The questions also give you a chance to practice your skills. If you have trouble with a question, talk with a parent or teacher.

Read each question carefully. If you do not know an answer, you may skip the question and come back to it later.

When you finish, check your answers.

Directions for Reading Assessment Two

Today you will take the Reading Assessment Two. You will read passages and answer questions. You may look back at the passage when you are answering the questions.

Directions to the Student

There are three different types of questions on this assessment:

1. There are multiple-choice questions that require you to choose the best answer.
2. There are short-answer questions for which you will write phrases or sentences on the lines provided in your booklet.
3. There are some extended-response questions for which you are expected to write a longer and more detailed answer in your booklet.

Here are some important things to remember as you take this assessment:

1. Read each passage. You may look back at the reading passage as often as you want.
2. The paragraphs are numbered for all reading passages. A question about a particular paragraph will refer to the paragraph number.
3. Read each question carefully. When you write your answers, write them neatly and clearly on the lines provided. You may use sentences, phrases, paragraphs, lists, or charts to explain your ideas. Cross out or erase any part of your work you do not want to include as part of your answer.
4. When you choose a multiple-choice answer, make sure you completely fill in the circle next to the answer. Erase completely any marks that you want to change on multiple-choice items.
5. Use only a **No. 2 pencil**, not a mechanical pencil or pen, to write your answers. If you do not have a No. 2 pencil, ask your teacher to give you one.
6. You should have plenty of time to finish every question on the assessment. If you do not know the answer to a question, go on to the next question. You can come back to that question later.
7. When you reach the word **STOP** in your booklet, you have reached the end of the Reading Assessment Two. Do **not** go on until you are told to turn the page.
8. If you finish early, you may check your work in this section **only**.

Go on ▶

Directions: Read the story and answer the questions.

from The Adventures of Robin Hood
by Howard Pyle

1   And so he came to dwell in the greenwood that was to be his home for many a year to come, never again to see the happy days with the lads and lasses of sweet Locksley Town; for he was outlawed, not only because he had killed a man, but also because he had poached upon the King's deer, and two hundred pounds were set upon his head, as a reward for whoever would bring him to the court of the King.

2   Now the Sheriff of Nottingham swore that he himself would bring this knave Robin Hood to justice, and for two reasons: first, because he wanted the two hundred pounds, and next, because the forester that Robin Hood had killed was of kin[1] to him.

3   But Robin Hood lay hidden in Sherwood Forest for one year, and in that time there gathered around him many others like himself, cast out from other folk for this cause and for that. Some had shot deer in hungry wintertime, when they could get no other food, and had been seen in the act by the foresters, but had escaped, thus saving their ears; some had been turned out of their inheritance, that their farms might be added to the King's lands in Sherwood Forest; some had been despoiled by a great baron or a rich abbot or a powerful esquire—all, for one cause or another, had come to Sherwood to escape wrong and oppression.

4   So, in all that year, fivescore or more good stout yeomen gathered about Robin Hood, and chose him to be their leader and chief. Then they vowed that even as they themselves had been despoiled they would despoil their oppressors, whether baron, abbot, knight, or squire, and that from each they would take that which had been wrung from the poor by unjust taxes, or land rents, or in wrongful fines. But to the poor folk they would give a helping hand in need and trouble, and would return to them that which had been unjustly taken from them. Besides this, they swore never to harm a

---

[1] *kin:* somebody related by blood; a relative

Go on ▶

child nor to wrong a woman, be she maid, wife, or widow; so that, after a while, when the people began to find that no harm was meant to them, but that money or food came in time of want to many a poor family, they came to praise Robin and his merry men, and to tell many tales of him and of his doings in Sherwood Forest, for they felt him to be one of themselves.

5   Up rose Robin Hood one merry morn when all the birds were singing blithely among the leaves, and up rose all his merry men, each fellow washing his head and hands in the cold brown brook that leaped laughing from stone to stone. Then said Robin, "For fourteen days have we seen no sport, so now I will go abroad to seek adventures forthwith. But tarry ye, my merry men all, here in the greenwood; only see that ye mind well my call. Three blasts upon the bugle horn I will blow in my hour of need; then come quickly, for I shall want your aid."

6   So saying, he strode away through the leafy forest glades until he had come to the verge of Sherwood. There he wandered for a long time, through highway and byway, through dingly dell and forest skirts. Now he met a fair buxom lass in a shady lane, and each gave the other a merry word and passed their way; now he saw a fair lady upon an ambling pad, to whom he doffed his cap, and who bowed sedately in return to the fair youth; now he saw a fat monk on a pannier-laden ass; now a gallant knight, with spear and shield and armor that flashed brightly in the sunlight; now a page clad in crimson; and now a stout burgher[2] from good Nottingham Town, pacing along with serious footsteps; all these sights he saw, but adventure found he none. At last he took a road by the forest skirts, a bypath that dipped toward a broad, pebbly stream spanned by a narrow bridge made of a log of wood. As he drew nigh this bridge he saw a tall stranger coming from the other side. Thereupon Robin quickened his pace, as did the stranger likewise, each thinking to cross first.

---

[2]*burgher: a prosperous citizen*

Go on ▶

1 How does the Sheriff of Nottingham feel about Robin Hood?

    ○ **A.** He is upset that Robin Hood has been outlawed.

    ○ **B.** He is angry with Robin Hood for stealing two hundred pounds.

    ○ **C.** He is sad because Robin Hood killed one of his deer.

    ○ **D.** He is angry that Robin Hood killed one of his relatives.

Go on ▶

2 How did Robin Hood's exile into Sherwood Forest influence him to become a hero to the poor? Include **three** details from the story in your answer.

Go on▶

3 What conclusion can the reader draw about Robin Hood? Provide information from the story to support your conclusion.

_____

_____

_____

_____

_____

_____

_____

_____

4 Which sentence best summarizes this story?

○ **A.** Robin Hood was outlawed from Nottingham and spent a year forming a group of people who wanted to rob the rich to help the poor.

○ **B.** Robin Hood was outlawed from Nottingham and spent a year alone in Sherwood Forest.

○ **C.** Robin Hood spent a year in Sherwood Forest and took money from rich people so he could live without working.

○ **D.** Robin Hood lived in Sherwood Forest, but ventured out occasionally to look for people to become friends with.

Go on ▶

5 Which word could the author have used in paragraph 4 instead of *despoiled*?

- A. Injured
- B. Robbed
- C. Offended
- D. Snubbed

6 What is the author's purpose for writing this story?

- A. To entertain readers with a fun story
- B. To inform readers about outlaws
- C. To describe to readers what life is like in a forest
- D. To explain how to make friends

Go on ▶

Directions: Read the selection and answer the questions.

Our National Bird

1   The bald eagle was chosen as the national symbol of the United States in 1782. It was chosen because of its long life, great strength, and majestic looks. At that time, people also believed that the bald eagle existed only in the area known as the United States. The eagle became a national emblem with the adoption of the Great Seal of the United States. The bald eagle serves as a sign of greatness and represents freedom. You can see the eagle on some U.S. coins, such as the quarter, the silver dollar, and the half dollar.

2   Eagles are large and beautiful birds with long, broad wingspans—from 70–90 inches. This helps them soar high in the air. Adult eagles have blackish-brown feathers on their bodies and white heads, necks, and tails. Some believe the white feathers on the eagle's head give it the appearance of being bald. At one time the word "bald" meant "white." However, the eagle really isn't bald or featherless at all.

3   Eagles are at the top of the food chain and have very few enemies. The eagle's powerful beak can tear through the prey it catches with its strong legs and talons. A part of the sea fish eagle group, these birds will eat both freshwater and saltwater fish. Eagles are known for their excellent eyesight and are able to see fish in the water from several hundred feet above. When they see a fish in the water, they will swoop down and catch it with their sharp talons or claws. Although they are mainly fish eaters, they will also feast on other available food sources, including dead animals. This fact adds to the eagle's reputation as a scavenger.

4   In the wild, eagles can live to be twenty or thirty years old. They are found in most of North America, ranging from Alaska and Canada to northern Mexico. The bald eagle was put on the endangered species list in 1967, because the birds were dying out. The birds have since made a reappearance, however. Today, there are about 50,000 bald eagles in the United States, and they are no longer considered an endangered species.

Go on ▶

7 Which sentence best explains how the bald eagle received its name?

- ○ A. The head of the bald eagle is featherless, but the rest of the bird has feathers.
- ○ B. The word "bald" once meant "white," and the bald eagle has white feathers on its head.
- ○ C. The bald eagle is a featherless bird.
- ○ D. The word "white" once meant "bald," and the bald eagle's head is featherless.

8 What is most likely the author's purpose for writing this selection?

- ○ A. To entertain the reader with stories about eagles.
- ○ B. To inform the reader about the importance of the bald eagle.
- ○ C. To persuade the reader to become involved with endangered species groups.
- ○ D. To describe to the reader what the Great Seal of the United States looks like.

Go on ▶

9 According to the selection, what happened when the bald eagle was designated as an endangered species?

- ○ A. Eagles only had a life span of 20 to 30 years.
- ○ B. Eagles were able to make a reappearance and they are no longer considered an endangered species.
- ○ C. Eagles moved to the top of the food chain.
- ○ D. Eagles became a national emblem with the adoption of the Great Seal of the United States.

10 Which word could the author have used in paragraph 1 instead of *majestic*?

- ○ A. Unrefined
- ○ B. Humble
- ○ C. Custom
- ○ D. Grand

Go on ▶

Directions: Read the fact box and answer the questions.

The Great Seal

The Great Seal of the United States shows a wide-spread bald eagle. On the eagle's breast, a shield appears. The shield contains 13 parallel red and white stripes. Also present is a blue field with the 13 stars. In the eagle's right talon, there is an olive branch. The left talon carries a bundle of thirteen arrows. In the eagle's beak, the bird carries a scroll inscribed with the motto: *e pluribus unum*, which translates as "one out of many."

Go on ▶

11 Explain **two** ways that the fact box "The Great Seal" helps you understand the selection. Include information from "The Great Seal" in your answer.

_____

_____

_____

_____

_____

_____

_____

_____

_____

12 What is the main similarity between "Our National Bird" and "The Great Seal"?

○ A. The selection and the fact box give information about the bald eagle and it's American symbolism.

○ B. The selection and the fact box show graphics of The Great Seal of the United States.

○ C. The selection and the fact box describe the life span and habitats of the bald eagle.

○ D. The selection and the fact box entertain the reader with stories about the bald eagle in the United States.

Go on ▶

13 What are **two** differences between the selection "Our National Bird" and the fact box "The Great Seal"? Include information from the selection in your answer.

_____

_____

_____

_____

_____

_____

_____

_____

Go on ▶

Directions: Read the selection and answer the questions.

The Iditarod

1   Each year in March, Alaska is host to the Iditarod Trail Sled Dog Races, where drivers from around the world compete. In this contest, drivers (or "mushers") and their teams of dogs race from Anchorage, in south central Alaska, to Nome, on the west Bering Sea coast. The Iditarod is often called "The Last Great Race on Earth."

2   An Alaskan woman named Dorothy Page thought of the idea for the Iditarod. Page chaired a committee that was looking for projects to celebrate Alaska's centennial year (the year marking 100 years of being a state): 1967. She wanted to honor the mark sled-dog teams had made on Alaska's history. In the years before airplane travel was common and before snowmobiles were in use, dog teams were the main method of transportation during the winter months. Two short races were held along parts of the Iditarod Trail in 1967 and 1969, but the first full-length race wasn't held until 1973.

3   In addition to raising people's awareness of the effect of dog teams on Alaska's history, the Iditarod celebrates two important parts of Alaska's history. The first is to honor the heroism and the feats of the 1925 mushers who relayed medicine almost 700 miles to save sick residents in Nome. The Great Race also celebrates the Iditarod National Historic Trail, which was one of Alaska's major mail routes.

4   The Iditarod is a challenging race. The race covers roughly 1,150 miles across Alaska's sometimes harsh and dangerous wilderness. Some of the contributing factors involve the uncertain weather conditions of the Alaskan countryside and the difficult land. Mushers are likely to face blinding snowstorms and temperatures below zero. They may also face other factors such as dangerous wild animals and unsafe hills. It takes great willpower, strength, and courage to complete such a task. No one can ever be sure how long it will take the winner to cross the finish line, but the race usually lasts 10 to 17 days.

Go on▶

14 What is the meaning of the word *harsh* as it is used in paragraph 4 of the selection?

○ A. Unknown

○ B. Rough

○ C. Hateful

○ D. Disagreeable

15 Based on the information in the selection, what assumption can the reader make about a future Iditarod race?

○ A. It will probably last longer than a week.

○ B. It will probably last longer than a month.

○ C. Mushers will probably face only good weather conditions.

○ D. It will be an easy task for the mushers to complete.

Go on ▶

16 Explain why the Iditarod is important to Alaskans. Include **two** details from the selection in your answer.

_____

_____

_____

_____

_____

_____

_____

_____

17 What is most likely the author's purpose for writing this selection?

○ **A.** To inform the reader about the importance and history of the Iditarod in Alaska

○ **B.** To explain the history of the Iditarod National Historic Trail to the reader

○ **C.** To raise awareness of the dog teams that race in the Iditarod to the reader

○ **D.** To entertain the reader with heroic stories about Alaskan dog teams

Go on ▶

18 Each of these titles could be another title for the selection. Choose the title you think best fits the selection.

The Amazing Race

A Journey for the Ages

An Icy Challenge

Provide **two** details from the selection to support your choice.

_____
_____
_____
_____
_____
_____
_____
_____

Go on ▶

Directions: Read the story and answer the questions.

The Professional

1   It was Marco's first job. The moment Mrs. Katz offered to pay him to take care of Savannah, he became a professional. He was only 12 years old.

2   Savannah, a large and beautiful female, already knew Marco. She ran out to meet him on the curb whenever he passed by her house. She always greeted him with his favorite sound: she sat down, she looked up at him, and she barked.

3   In February, a week after Marco's birthday, Mrs. Katz took a trip out of town. She left Savannah in Marco's care. His job was to stop by twice a day to feed and walk Savannah. Behind a tall basket on the porch, Mrs. Katz left her house key and Savannah's leather leash. The salary was great—four dollars per visit—and all Marco had to do was play with his favorite pet. Mrs. Katz even printed up a business card that included Marco's phone number and the title "Dog Sitter."

4   On Monday, the first day of the job, Marco awoke before dawn. In a flash, he put on warm clothing and pulled on snow boots. Cold air blew inside his collar as he rolled his bike out of the shed. A new layer of midnight snow covered the icy road.

5   "This is definitely the only bicycle on the street after a four-inch snow fall," he thought. At least the sidewalk had been cleared. He felt mature and responsible. No one else in the neighborhood seemed to be awake, yet he was already out of bed, dressed, and on his way to work.

6   Mrs. Katz's house was two blocks down the street. Her porch seemed to reach out over the top of the hill. When Marco let himself in at the back door, he found everything exactly where Mrs. Katz had said it would be. Running toward him from the silence of the empty house, Savannah gave him a three-minute hello. She was definitely glad to see him.

7   First, he tackled the kitchen duties. He cleaned out the water bowl, filled it with fresh water, and poured a scoop of dog food into Savannah's food bowl. One swipe with a paper towel was enough to give the floor a tidy appearance.

8   When he opened the kitchen door, Savannah leapt past him, out toward the snow. Marco held out the leash to show it was time for a walk. Savannah ran over obediently. Marco loved this part. She felt like his dog whenever he walked her on the leash.

Go on ▶

9   Together, they trudged across the new snow. The morning seemed to shimmer and sparkle. Savannah's long, thin legs sank into deep snowdrifts with a crunchy-sounding step. When they reached Marco's house, at the bottom of the hillside, they turned to go back up the street. Slipping and sliding up the incline, the boy and the dog ran toward Mrs. Katz's house.

10 Marco's homemade chart rested on the kitchen table. He had carefully drawn a system for recording each visit and each feeding time. At the end of this first visit, he noted his time on the chart. For amusement, he also recorded Savannah's mood. This would be a fun feature of the job because Savannah was always in a good mood.

11 Before leaving, Marco stood proudly in Mrs. Katz's living room. Looking into the tall, elegant mirror, he smiled at the professional who was looking back at him.

Go on▶

19 Which sentence explains why Marco and Savannah were slipping and sliding up the street?

- A. The street was icy and cold.
- B. The street was dry and warm.
- C. The street was wet and muddy.
- D. It was raining outside.

20 What is the meaning of the word *tackled* as it is used in paragraph 7 of the story?

- A. To harness
- B. To take hold of
- C. To undertake
- D. To deal with

Go on ▶

21 What does the author mean when he/she says, "the morning seemed to shimmer and sparkle" in paragraph 9 of the story?

○ **A.** The sky looked shiny and bright.

○ **B.** New snow made the morning appear to be bright and glisten.

○ **C.** The sun was shining too bright.

○ **D.** Marco was so happy that the morning sparkled.

22 Explain why Marco felt like a professional. Include **two** details from the story in your answer.

_____

_____

_____

_____

_____

_____

_____

_____

Go on▶

Directions: Read the selection and answer the questions.

Cozumel

An Island Paradise

1   What if you could visit a place that offers clear blue water, crystal beaches, and underwater adventures? Cozumel, in Mexico, is just such a place. Located off Mexico's Yucatan Peninsula, Cozumel has become a favorite vacation spot for divers and those who love the outdoors.

2   This exciting island is only 10 miles long and 28 miles wide, but it is Mexico's largest island. It attracts many visitors from across the globe each year. Most people travel to Cozumel to experience underwater diving at its best. It is thought of as one of the best diving spots in the world, second only to Australia.

3   Snorkeling is another activity Cozumel visitors enjoy. It is the second most popular water sport on the island. Beginners can snorkel right off the beaches of the hotels and see a variety of fish without going out too far. Another good place for beginning snorkelers is Chankanaab National Park. This park has shallow reefs that attract fish of all shapes, colors, and sizes.

4   If you're not interested in water sports, the Chankanaab Park also has nature trails where visitors can spend time walking and admiring the different plants of the island. You might even spot an iguana or two taking an afternoon nap in the sun!

5   San Miguel is the only town in Cozumel. Tourists from cruise ships and others who make the trip to town are in for a real treat. There are plenty of people who live in the town selling T-shirts, silver jewelry, and other goods made in Mexico.

6   It's easy to see why Cozumel is a favorite choice for outdoor adventures and sun-filled excitement!

Go on ▶

Beyond the Vacation

7   Nestled between the trade routes to and from Honduras and Veracruz, Cozumel was well positioned as a seaport. Settled as early as 300 A.D. by the Maya, this island served many purposes. These purposes included being a pilgrimage site (a sacred place people journey to) and the center for Mayan trade.

8   As the Spanish arrived in the area in the late 1400s and early 1500s, the Maya resisted a number of attempts at Spanish settlement. By 1519, however, a bitter struggle for the Yucatan Peninsula began. Slowly, the struggle crept outward. A Spanish presence began to take over the island. With the arrival of Hernan Cortéz and his men, many of the Mayan temples and shrines were destroyed. The foreigners also exposed the natives to smallpox. An epidemic (a very fast and wide spread illness) broke out, and by 1570, Cozumel's population declined to fewer than 300 people.

9   Throughout the 17th century, Cozumel was occupied mainly by pirates. The location of the island provided privacy and protection from danger. The area wasn't re-inhabited until 1848. At that time, Spanish settlers sought refuge from the Caste War, which was being fought on the mainland. A quiet fishing village was eventually established, and Cozumel remained as such until 1961. In that year, French explorer Jacques Cousteau declared the area's waters to be some of the most amazing for exploration. Since then, Cozumel has become a popular tourist destination, welcoming visitors from around the world every year.

Go on▶

23 Which sentence tells how the two parts of the selection are different?

- A. "An Island Paradise" talks about the history of Cozumel; "Beyond the Vacation" talks about the island in the present.
- B. "An Island Paradise" talks about Cozumel in the present; "Beyond the Vacation" talks about the history of the island.
- C. "An Island Paradise" talks about Cozumel in the present; "Beyond the Vacation" talks about the future of the island.
- D. "An Island Paradise" talks about Cozumel in the future; "Beyond the Vacation" talks about the history of the island.

24 What is the meaning of the word *occupied* as it is used in paragraph 9 of the selection?

- A. To do business with
- B. To take up space
- C. To take possession of
- D. To dwell or reside

25 According to the selection, what happened when foreigners exposed the natives of Cozumel to smallpox?

- A. Cozumel has become a popular tourist destination.
- B. Cozumel was well positioned as a seaport.
- C. Cozumel's population declined to fewer than 300 people.
- D. Many of the Mayan temples and shrines were destroyed.

Go on▶

26 What is the author's purpose for writing the selection? Provide **three** details from the selection to support your answer.

27 What are **two** similarities between "An Island Paradise" and "Beyond the Vacation"? Include information from the selection in your answer.

_____

_____

_____

_____

_____

_____

_____

_____

_____

Go on ▶

Directions: Read the story and answer the questions.

Stormy Nights

1   Hendrik still felt afraid of storms at night. If he awoke during a thunderstorm, fear seemed to take a grip on his throat until he could barely breathe.

2   Hendrik's bedroom looked ghostly when a storm shook the neighborhood. The room suddenly appeared white during flashes of lightning and then instantly grew darker-than-night during crashes of thunder. The flashes consisted of an unearthly light that shone in fits and starts: it turned on, then off, then on again. It made the room appear to move.

3   Imagine a bedroom rocking like a ship at sea! Hendrik would lie perfectly still under his warm, neat sheets, yet he felt as though he was drowning.

4   By the middle of June, there seemed to be a fresh storm every few nights. Often, the rain fell after midnight. By morning, things would look beautiful and damp. At dawn, the June grass lay like a smooth, green carpet glistening across the world. He tried to picture this at night, but the grass seemed different in darkness. After sunset, nothing looked green or smooth.

5   On the last night in June, Hendrik made himself comfortable on the old porch swing. He sat for a while. The night was completely dark, and it was impossible to see the dark grass. Silently, Hendrik watched and listened. He could hear life in the grass: thousands of crickets were communicating in the night. He thought about these little creatures that spend their entire lives outdoors, drinking the rainwater and the dew they discover on blades of grass. "It's amazing how those tiny insects survive all these turbulent storms," he thought. The sound of their chirping washed around him like gentle waves. The sound was comforting.

Go on

28 What does the author mean when the author says, "The flashes consisted of an unearthly light that shone in fits and starts" in paragraph 2 of the story?

- A. Hendrik thought he saw a flash of light, but it was only his imagination.
- B. The light appeared, then disappeared, then reappeared.
- C. The light disappeared.
- D. Hendrik enjoyed the light as it shined brightly through the night.

29 What is the meaning of the word *glistening* as it is used in paragraph 4 of the story?

- A. Giving a dull appearance
- B. Giving a shiny appearance
- C. Giving a green appearance
- D. Giving a dry appearance

30 Which two objects does the author compare using a simile?

- A. Lightening to crashing thunder
- B. Night to complete darkness
- C. Bedroom to a rocking ship
- D. Rainwater to the morning dew

Go on ▶

31 What is the author's purpose for writing the story?

- A. To describe the science behind thunderstorms
- B. To explain why thunderstorms scare kids
- C. To entertain readers with a story about the a boy's fear of thunderstorms
- D. To inform readers how thunderstorms form

32 Based on the information in the story, which conclusion can the reader draw about thunderstorms?

- A. Thunderstorms are scary and not useful.
- B. Thunderstorms can be scary but are important.
- C. Thunderstorms cause a lot of damage.
- D. Thunderstorms disrupt the balance of nature.

33 Which sentence best summarizes the story?

- A. Hendrik was afraid of thunderstorms, but realized that they were beneficial.
- B. Hendrik was afraid of thunderstorms and could not sleep through them.
- C. Hendrik spent his summer nights watching thunderstorms from the front porch.
- D. Hendrik listened for the sounds of crickets on nights when there were no thunderstorms.

STOP

Mathematics

Introduction

In the Mathematics Assessment of the Measurements of Student Progress (MSP), you will be asked questions to test the knowledge you have learned so far in school. These questions are based on the mathematical skills you have been taught in school through sixth grade. The questions you answer are not meant to confuse or trick you, but are written so you have the best opportunity to show what you know about mathematics.

The *Show What You Know® on the MSP for Grade 6, Student Workbook* includes a Mathematics Tutorial that will help you practice your test-taking skills. Following the Mathematics Tutorial is a full-length Mathematics Assessment. Both the Mathematics Tutorial and the Mathematics Assessment have been created to model the Grade 6 Measurements of Student Progress for Mathematics.

About the Mathematics MSP

The Grade 6 Mathematics Assessment will test Content (numbers, operations, algebra, geometry/measurement, data/statistics/probability) as well as Process (reasoning, problem solving, and communication). The Mathematics Assessment is given in one session.

For the Mathematics Assessment there are four different types of questions: multiple choice, completion, short answer, and extended response. Dictionaries, thesauruses, and scratch paper are not allowed on the Mathematics Assessment.

Scoring

On the MSP for Grade 6 Mathematics Assessment, each multiple-choice item is worth one point. Short-answer items will be scored on a scale of zero to two points. Extended-response items will be scored on a scale of zero to four points. The scoring criteria will focus on the understanding of mathematical ideas, information, and solutions, and will disregard conventions of writing (complete sentences, usage/grammar, spelling, capitals, punctuation, and paragraphing), as long as the wording of the response does not interfere with the mathematical communication.

Typical Distribution of Score Points by Item Type*

Type	Number of Items	Total Possible Points
Multiple Choice	26	26
Short Answer	8	16
Extended Response	2	8
Total	36	50

*2009 testing information

Glossary

addend: Numbers added together to give a sum. For example, 2 + 7 = 9. The numbers 2 and 7 are addends.

addition: An operation joining two or more sets where the result is the whole.

a.m.: The hours from midnight to noon; from Latin words *ante meridiem* meaning "before noon."

analyze: To break down information into parts so that it may be more easily understood.

angle: A figure formed by two rays that meet at the same end point called a vertex. Angles can be obtuse, acute, right, or straight.

area: The number of square units needed to cover a region. The most common abbreviation for area is *A*.

Associative Property of Addition: The grouping of addends can be changed and the sum will be the same. Example: (3 + 1) + 2 = 6; 3 + (1 + 2) = 6.

Associative Property of Multiplication: The grouping of factors can be changed and the product will be the same. Example: (3 x 2) x 4 = 24; 3 x (2 x 4) = 24.

attribute: A characteristic or distinctive feature.

average: A number found by adding two or more quantities together and then dividing the sum by the number of quantities. For example, in the set {9, 5, 3}, the average is 6: 9 + 5 + 4 = 18; 18 ÷ 3 = 6. *See mean.*

axes: Plural of axis. Perpendicular lines used as reference lines in a coordinate system or graph; traditionally, the horizontal axis (*x*-axis) represents the independent variable and the vertical axis (*y*-axis) represents the dependent variable.

bar graph: A graph using bars to show data.

capacity: The amount an object holds when filled.

chart: A way to show information, such as in a graph or table.

circle: A closed, curved line made up of points that are all the same distance from a point inside called the center. Example: A circle with center point *P* is shown below.

circle graph: Sometimes called a pie chart; a way of representing data that shows the fractional part or percentage of an overall set as an appropriately sized wedge of a circle. Example:

circumference: The boundary line or perimeter of a circle; also, the length of the perimeter of a circle. Example:

Commutative Property of Addition: Numbers can be added in any order and the sum will be the same. Example: 3 + 4 = 4 + 3.

Commutative Property of Multiplication: Numbers can be multiplied in any order and the product will be the same. Example: 3 x 6 = 6 x 3.

compare: To look for similarities and differences. For example, is one number greater than, less than, or equal to another number?

conclusion: A statement that follows logically from other facts.

Glossary

cone: A solid figure with a circle as its base and a curved surface that meets at a point.

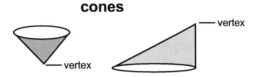
cones

congruent figures: Figures that have the same shape and size.

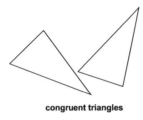
congruent triangles

cube: A solid figure with six faces that are congruent (equal) squares.

cylinder: A solid figure with two circular bases that are congruent (equal) and parallel to each other connected by a curved lateral surface.

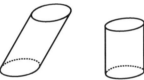

data: Information that is collected.

decimal number: A number expressed in base 10, such as 39.356, where each digit's value is determined by multiplying it by some power of 10.

denominator: The bottom number in a fraction.

diagram: A drawing that represents a mathematical situation.

difference: The answer when subtracting two numbers.

distance: The amount of space between two points.

dividend: A number in a division problem that is divided. Dividend ÷ divisor = quotient. Example: In 15 ÷ 3 = 5, 15 is the dividend.

$$\text{divisor} \overline{)\text{dividend}}^{\text{quotient}} \qquad 3\overline{)15}^{\,5}$$

divisible: Can be divided by another number without leaving a remainder. Example: 12 is divisible by 3 because 12 ÷ 3 is an integer, namely 4.

division: An operation that tells how many equal groups there are or how many are in each group.

divisor: The number by which another number is divided. Example: In 15 ÷ 3 = 5, 3 is the divisor.

$$\text{divisor} \overline{)\text{dividend}}^{\text{quotient}} \qquad 3\overline{)15}^{\,5}$$

edge: The line segment where two faces of a solid figure meet.

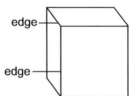

equality: Two or more sets of values that are equal.

equation: A number sentence that says two expressions are equal (=). Example: 4 + 8 = 6 + 6.

equivalent fractions: Two fractions with equal values.

estimate: To find an approximate value or measurement of something without exact calculation.

even number: A whole number that has a 0, 2, 4, 6, or 8 in the ones place. A number that is a multiple of 2. Examples: 0, 4, and 678 are even numbers.

expanded form: A number written as the sum of the values of its digits. Example: 546 = 500 + 40 + 6.

Glossary

expression: A combination of variables, numbers, and symbols that represent a mathematical relationship.

face: The sides of a solid figure. For example, a cube has six faces that are all squares. The pyramid below has five faces—four triangles and one square.

fact family: A group of related facts using the same numbers. Example: 5 + 8 = 13; 13 − 8 = 5.

factor: One of two or more numbers that are multiplied together to give a product. Example: In 3 x 4 = 12, 3 and 4 are factors of 12.

figure: A geometric figure is a set of points and/or lines in 2 or 3 dimensions.

flip (reflection): The change in a position of a figure that is the result of picking it up and turning it over. Example: Reversing a "b" to a "d."
Tipping a "p" to a "b" or a "b" to a "p" as shown below:

fraction: A symbol, such as $\frac{2}{8}$ or $\frac{5}{3}$, used to name a part of a whole, a part of a set, or a location on the number line. Examples:

$$\frac{\text{numerator}}{\text{denominator}} = \frac{\text{dividend}}{\text{divisor}}$$

$$\frac{\text{\# of parts under consideration}}{\text{\# of parts in a set}}$$

function machine: Applies a function rule to a set of numbers, which determines a corresponding set of numbers.
Example: Input 9 → Rule x 7 → Output 63. If you apply the function rule "multiply by 7" to the values 5, 7, and 9, the corresponding values are:
$$5 \rightarrow 35$$
$$7 \rightarrow 49$$
$$9 \rightarrow 63$$

graph: A "picture" showing how certain facts are related to each other or how they compare to one another. Some examples of types of graphs are line graphs, pie charts, bar graphs, scatterplots, and pictographs.

grid: A pattern of regularly spaced horizontal and vertical lines on a plane that can be used to locate points and graph equations.

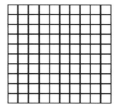

hexagon: A six-sided polygon. The total measure of the angles within a hexagon is 720°.

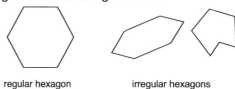

regular hexagon   irregular hexagons

impossible event: An event that can never happen.

integer: Any number, positive or negative, that is a whole number distance away from zero on a number line, in addition to zero. Specifically, an integer is any number in the set {. . .-3,-2,-1, 0, 1, 2, 3. . .}. Examples of integers include: 1, 5, 273, -2, -35, and -1,375.

intersecting lines: Lines that cross at a point. Examples:

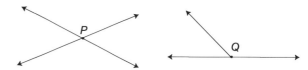

Glossary

isosceles triangle: A triangle with at least two sides the same length.

justify: To prove or show to be true or valid using logic and/or evidence.

key: An explanation of what each symbol represents in a pictograph.

kilometer (km): A metric unit of length: 1 kilometer = 1,000 meters.

line: A straight path of points that goes on forever in both directions.

line graph: A graph that uses a line or a curve to show how data changes over time.

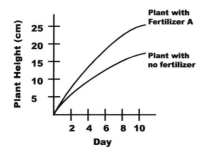

line of symmetry: A line on which a figure can be folded into two parts so that the parts match exactly.

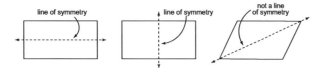

liter (L): A metric unit of capacity: 1 liter = 1,000 milliliters.

mass: The amount of matter an object has.

mean: Also called arithmetic average. A number found by adding two or more quantities together, and then dividing the sum by the number of quantities. For example, in the set {9, 5, 3} the mean is 6: 9 + 5 + 4 = 18; 18 ÷ 3 = 6. *See average.*

median: The middle number when numbers are put in order from least to greatest or from greatest to least. For example, in the set of numbers 6, 7, 8, 9, 10, the number 8 is the median (middle) number.

meter (m): A metric unit of length: 1 meter = 100 centimeters.

method: A systematic way of accomplishing a task.

mixed number: A number consisting of a whole number and a fraction.
Example: $6\frac{2}{3}$.

mode: The number or numbers that occur most often in a set of data. Example: The mode of {1, 3, 4, 5, 5, 7, 9} is 5.

multiple: A product of a number and any other whole number. Examples: {2, 4, 6, 8, 10, 12,...} are multiples of 2.

multiplication: An operation on two numbers that tells how many in all. The first number is the number of sets and the second number tells how many in each set.

number line: A line that shows numbers in order using a scale. Equal intervals are marked and usually labeled on the number line.

number sentence: An expression of a relationship between quantities as an equation or an inequality. Examples: 7 + 7 = 8 + 6; 13 < 92; 56 + 4 > 59.

numerator: The top number in a fraction.

octagon: An eight-sided polygon. The total measure of the angles within an octagon is 1080°.

regular octagon   irregular octagons

odd number: A whole number that has 1, 3, 5, 7, or 9 in the ones place. An odd number is not divisible by 2. Examples: The numbers 53 and 701 are odd numbers.

operation: A mathematical process that combines numbers; basic operations of arithmetic include addition, subtraction, multiplication, and division.

order: To arrange numbers from the least to greatest or from the greatest to least.

Glossary

ordered pair: Two numbers inside a set of parentheses separated by a comma that are used to name a point on a coordinate grid.

parallel lines: Lines in the same plane that never intersect.

parallelogram: A quadrilateral in which opposite sides are parallel.

pattern: An arrangement of numbers, pictures, etc., in an organized and predictable way. Examples: 3, 6, 9, 12, or ® 0 ® 0 ® 0.

pentagon: A five-sided polygon. The total measure of the angles within a pentagon is 540°.

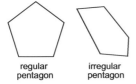

perimeter: The distance around a figure.

perpendicular lines: Two lines that intersect to form a right angle (90 degrees).

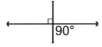

pictograph: A graph that uses pictures or symbols to represent similar data. The value of each picture is interpreted by a "key" or "legend."

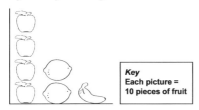

place value: The value given to the place a digit has in a number.
Example: In the number 135, the 1 is in the hundreds place so it represents 100 (1 x 100), the 3 is in the tens place so it represents 30 (3 x 10), and the 5 is in the ones place so it represents 5 (5 x 1).

p.m.: The hours from noon to midnight; from the Latin words *post meridiem* meaning "after noon."

point: An exact position often marked by a dot.

polygon: A closed figure made up of straight line segments.

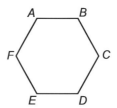

ABCDEF is a polygon.

possible event: An event that might or might not happen.

predict: To tell what you believe may happen in the future.

prediction: A description of what may happen before it happens.

probability: The likelihood that something will happen.

product: The answer to a multiplication problem. Example: In 3 x 4 = 12, 12 is the product.

pyramid: A solid figure in which the base is a polygon and faces are triangles with a common point called a vertex.

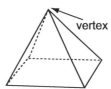

quadrilateral: A four-sided polygon. Rectangles, squares, parallelograms, rhombi, and trapezoids are all quadrilaterals. The total measure of the angles within a quadrilateral is 360°. Example: ABCD is a quadrilateral.

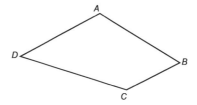

questionnaire: A set of questions for a survey.

Glossary

quotient: The answer in a division problem. Dividend ÷ divisor = quotient. Example: In 15 ÷ 3 = 5, 5 is the quotient.

range: The difference between the least number and the greatest number in a data set. For example, in the set {4, 7, 10, 12, 36, 7, 2}, the range is 34; the greatest number (36) minus the least number (2): (36 – 2 = 34).

rectangle: A quadrilateral with four right angles. A square is one example of a rectangle.

reflection: The change in the position of a figure that is the result of picking it up and turning it over. *See flip.*

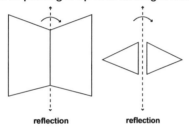

remainder: The number that is left over after dividing. Example: In 31 ÷ 7 = 4 R 3, the 3 is the remainder.

represent: To present clearly; describe; show.

rhombus: A quadrilateral with opposite sides parallel and all sides the same length. A square is one kind of rhombus.

right angle: An angle that forms a square corner and measures 90 degrees.

right triangle: A triangle having one right angle. *See angle and triangle.*

rounding: Replacing an exact number with a number that tells about how much or how many to the nearest ten, hundred, thousand, and so on. Example: 52 rounded to the nearest 10 is 50.

rule: A procedure; a prescribed method; a way of describing the relationship between two sets of numbers. Example: In the following data, the rule is to add 3:

Input	Output
3	6
5	8
9	12

ruler: A straight-edged instrument used for measuring the lengths of objects. A ruler usually measures smaller units of length, such as inches or centimeters.

scale: The numbers that show the size of the units used on a graph.

sequence: A set of numbers arranged in a special order or pattern.

set: A group made up of numbers, figures, or parts.

side: A line segment connected to other segments to form the boundary of a polygon.

similar: A description for figures that have the same shape.

slide (translation): The change in the position of a figure that moves up, down, or sideways. Example: scooting a book on a table.

solids: Figures in three dimensions.

solve: To find the solution to an equation or problem; finding the values of unknown variables that will make a true mathematical statement.

sphere: A solid figure in the shape of a ball. Example: a basketball is a sphere.

Glossary

square: A rectangle with congruent (equal) sides. *See rectangle.*

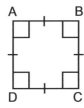

square number: The product of a number multiplied by itself. Example: 49 is a square number (7 x 7 = 49).

square unit: A square with sides 1 unit long, used to measure area.

standard form: A way to write a number showing only its digits. Example: 2,389.

standard units of measure: Units of measure commonly used; generally classified in the U.S. as the customary system or the metric system:

Customary System:
Length
1 foot (ft) = 12 inches (in)
1 yard (yd) = 3 feet or 36 inches
1 mile (mi) = 1,760 yards or 5,280 feet

Weight
16 ounces (oz) = 1 pound (lb)
2,000 pounds = 1 ton (t)

Capacity
1 pint (pt) = 2 cups (c)
1 quart (qt) = 2 pints
1 gallon (gal) = 4 quarts

Metric System:
Length
1 centimeter (cm) = 10 millimeters (mm)
1 decimeter (dm) = 10 centimeters
1 meter (m) = 100 centimeters
1 kilometer (km) = 1,000 meters

Weight
1,000 milligrams (mg) = 1 gram (g)
1,000 grams (g) = 1 kilogram (kg)
1,000 kilograms (kg) = 1 tonne (metric ton)

Capacity
1 liter (l) = 1,000 milliliters (ml)

---

strategy: A plan used in problem solving, such as looking for a pattern, drawing a diagram, working backward, etc.

subtraction: The operation that finds the difference between two numbers.

sum: The answer when adding two or more addends. Addend + Addend = Sum.

summary: A series of statements containing evidence, facts, and/or procedures that support a result.

survey: A way to collect data by asking a certain number of people the same question and recording their answers.

symmetry: A figure has line symmetry if it can be folded along a line so that both parts match exactly. A figure has radial or rotational symmetry if, after a rotation of less than 360°, it is indistinguishable from its former image.

Z

Z unrotated   Z rotated 90°   Z rotated 180°

The letter Z has 180° radial or rotational symmetry.

table: A method of displaying data in rows and columns.

temperature: A measure of hot or cold in degrees.

translation (slide): A change in the position of a figure that moves it up, down, or sideways.

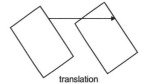

translation

Glossary

triangle: A polygon with three sides. The sum of the angles of a triangle is always equal to 180°.

turn: The change in the position of a figure that moves it around a point. Also called a rotation. Example: The hands of a clock turn around the center of the clock in a clockwise direction.

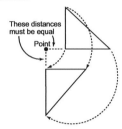

unlikely event: An event that probably will not happen.

vertex: The point where two rays meet to form an angle or where the sides of a polygon meet, or the point where 3 or more edges meet in a solid figure.

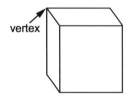

whole number: An integer in the set {0, 1, 2, 3 . . .}. In other words, a whole number is any number used when counting in addition to zero.

word forms: The number written in words. Examples: 546 is "five hundred forty-six."

Examples of Common Two-Dimensional Geometric Shapes

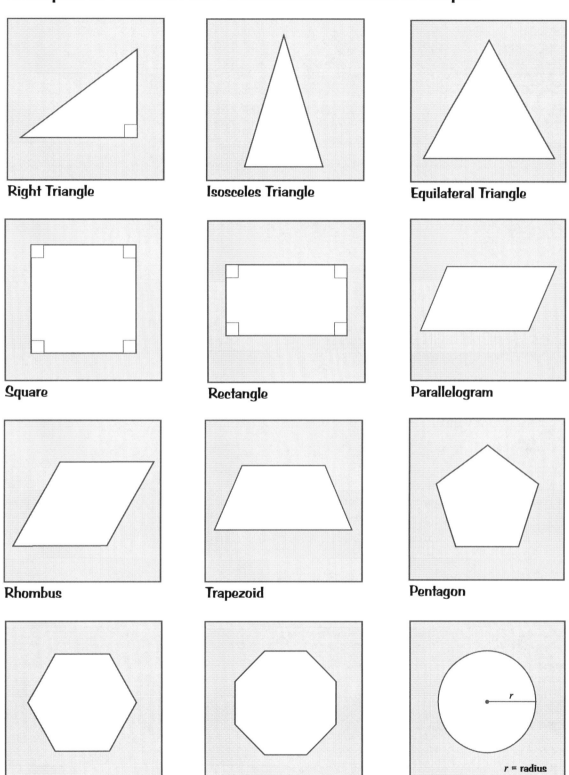

Examples of How Lines Interact

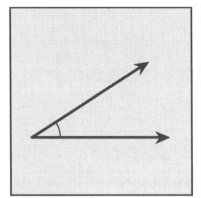
Acute Angle

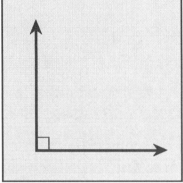
Right Angle

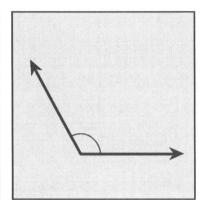
Obtuse Angle

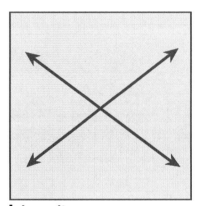
Intersecting

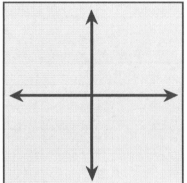
Perpendicular

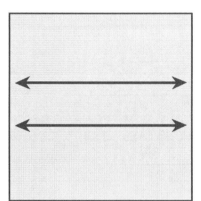
Parallel

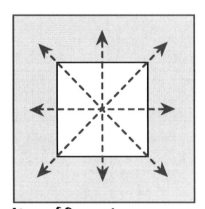
Lines of Symmetry

Examples of Common Types of Graphs

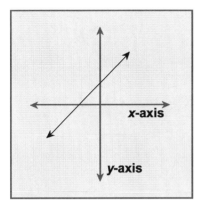
Line Graph

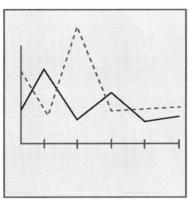
Double Line Graph

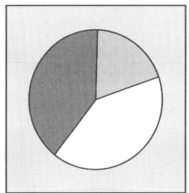
Pie Chart

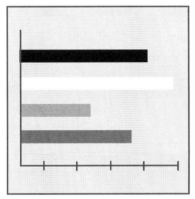
Bar Graph

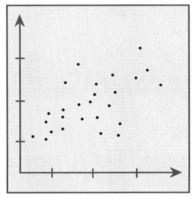
Scatterplot

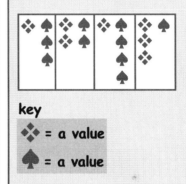
Pictograph

Examples of Common Three-Dimensional Objects

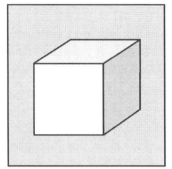
Cube

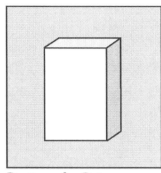
Rectangular Prism

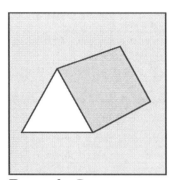
Triangular Prism

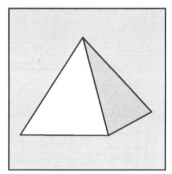
Pyramid

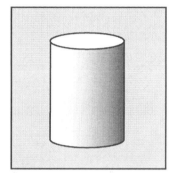
Cylinder

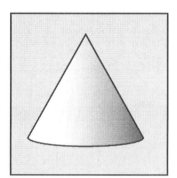
Cone

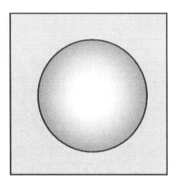
Sphere

Examples of Object Movement

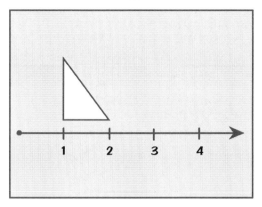 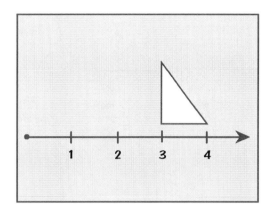

Translation

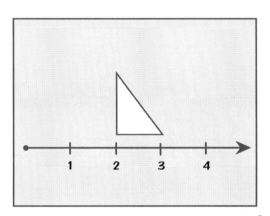 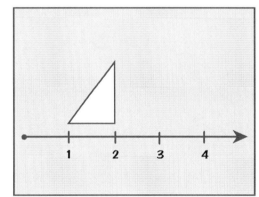

Reflection

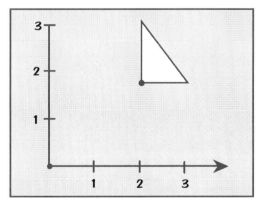 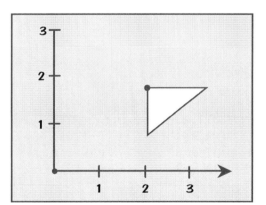

Rotation

This page left intentionally blank.

Mathematics Tutorial

The Mathematics Tutorial is made up of multiple-choice questions, completion items, short-answer, and extended-response questions. These questions show you how the skills you have learned in Mathematics class may be tested on the Mathematics MSP. The questions also give you a chance to practice your skills. If you have trouble with an area, talk with a parent or teacher.

Read each question carefully. If you do not know an answer, you may skip the question and come back to it later.

When you finish, check your answers.

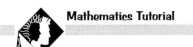

Mathematics Tutorial

Directions for Mathematics Tutorial

Today you will take the Mathematics Tutorial. On this part of the assessment, you are permitted to use tools such as calculators, rulers, protractors, or manipulatives.

Directions to the Student

There are several different types of questions on this assessment:

1. Some questions will ask you to choose the best answer from among four answer choices. These items are worth one point.

2. Some questions will ask you to write or draw an answer neatly and clearly **inside** an answer box.

    - Some of these questions are short. They may ask you to write an answer, to show how you got your answer using words, numbers, or pictures, or show the steps you used to solve the problem. These items are worth two points.

    - Others ask for more details (graphs, tables, written summaries). These questions also provide you with more room for your answer. These items are worth four points.

Here are some important things to remember as you take this assessment:

1. Read each question carefully and think about the answer.

2. When you choose a multiple-choice answer, make sure you completely fill in the circle next to the answer. Erase completely any marks that you want to change on multiple-choice items.

3. When an answer box is provided, write your answer neatly and clearly **inside** the box and show all your work. Cross out any work you do not want as part of your answer. **Do not use scratch paper.**

4. Use only a **No. 2 pencil**, not a mechanical pencil or pen, to write your answers. If you do not have a No. 2 pencil, ask your teacher to give you one.

5. You should have plenty of time to finish every question on the assessment. If you do not know the answer to a question, go on to the next question. You can come back to that question later.

6. When you reach the word **STOP** in your booklet, you have reached the end of the tutorial. Do **not** go on until you are told to turn the page.

7. If you finish early, you may check your work in this section **only**.

Go on ▶

Sample Questions

To help you understand how to answer the test questions, look at the sample test questions below. They are included to show you what the questions in the test are like and how to mark or write your answers.

Multiple-Choice Sample Question

For this type of question you will select the answer and fill in the circle next to it.

1 For his birthday, Pedro took 63 cookies to share with his class. There are a total of 21 students in the class.

   What is the ration of cookies to students?

   ● A. 3 to 1
   ○ B. 1 to 3
   ○ C. 7 to 1
   ○ D. 1 to 7

For this sample question, the correct answer is A; therefore, the circle next to A is filled in.

Completion Item Sample Question

For this type of question you will provide a short answer such as a single number or one or two words.

2 Milton was watching TV. He was watching Channel 10. During a commercial, he flipped 3 channels up. Then, he flipped 5 channels down. Finally, he flipped 7 channels up and 2 channels down.

   • Write what channel Milton ended on.

   What channel did Milton end on? __13__

Go on ▶

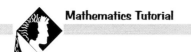

Short-Answer Sample Question

For this type of question you will write and explain your answer using words, numbers, or pictures.

3 Andy is 3 years younger than 2 times his sister's age, n.

Write an equation that can be used to find Andy's age, a.

Show your work using words, numbers, and/or pictures.

> Andy's age is 3 less (−3) than 2 times his sister's age (2n).
>
> $a = 2n - 3$

Go on ▶

Extended-Response Sample Question

For this type of question you will write an extended answer offering more examples and more detail. You may use words, numbers, and/or pictures.

4 The revenues for last year of the Pitzulo Spaghetti Sauce Company are shown on the graph below.

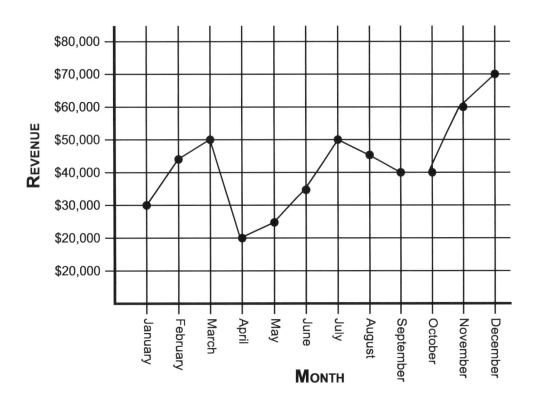

Go on ▶

Study the graph on page 131.

- Give **two** conclusions about the Pitzulo Spaghetti Sauce Company revenues.
- Support each conclusion using specific data from the graph.

Conclusion number 1 with support from the graph:

> The Pitzulo Spaghetti Sauce Company had the greatest increase in revenue from October to November. There was a $20,000 increase in revenue from October to November.

Conclusion number 2 with support from the graph:

> The Pitzulo Spaghetti Sauce Company had the greatest difference in revenue from March to April. There was a $30,000 decrease in revenue from March to April.

Go on ▶

1 Janie and her friends are comparing numbers on note cards they drew from a hat in math class. The teacher asked them to line up according to the relative values of their numbers. List the numbers $\frac{8}{5}$, $1\frac{2}{5}$, 1.09, and 1.65 on the number line.

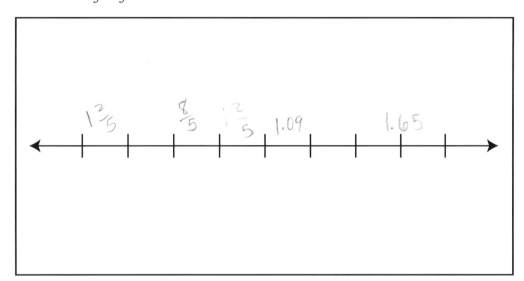

2 Beanie has eaten $\frac{2}{3}$ of a bag of popcorn and Jim has eaten $\frac{2}{5}$ of the amount Beanie has eaten.

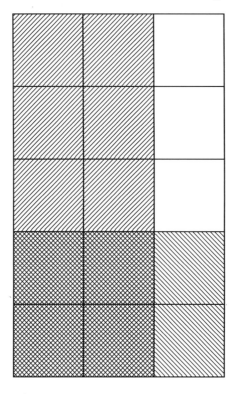

Which equation is pictured in the model and can be used to calculate how much popcorn Jim has eaten?

○ A. $\frac{2}{3} \div \frac{2}{5}$

● B. $\frac{2}{5} \times \frac{2}{3}$

○ C. $\frac{1}{3} \times \frac{2}{5}$

○ D. $\frac{2}{5} + \frac{2}{3}$

Go on▶

3 Mr. Bair wants to put fertilizer in his vegetable garden. Before he buys the bags of fertilizer, he wants to find out the size of his garden.

The dimensions of the garden are 0.48 yards × 16.1 yards.

What is a good estimate of the area of Mr. Bair's vegetable garden?

- A. 6 square yards
- ● B. 7 square yards
- C. 8 square yards
- D. 9 square yards

4 Mary wants to make postcards that have an area of $23\frac{3}{8}$ square inches. She knows that one side of the card will be $5\frac{1}{2}$ inches in length.

What will be the length of the other side of the card?

- ● A. $4\frac{1}{4}$ inches
- B. $4\frac{3}{5}$ inches
- C. 5 inches
- D. $5\frac{1}{4}$ inches

Go on ▶

5 Javier collects pennies. In his collection, he has 3,400 pennies.

How many dollars does Javier have in his collection?

- ○ A. $0.34
- ○ B. $3.40
- ● C. $34.00
- ○ D. $340.00

6 Drew is covering his garage floor with epoxy. He needs to calculate the area of his garage floor so he knows how much epoxy to purchase.

Drew's garage is 18.75 feet long and 16.2 feet wide.

What is the area of Drew's garage floor?

- ○ A. 30.375 square feet
- ○ B. 174.75 square feet
- ○ C. 247 square feet
- ● D. 303.75 square feet

Go on ▶

7 Without doing any computation, list 83, 0.56 × 83, and 83 ÷ 0.95 in increasing order and explain your reasoning.

8 A spelling test had 20 words. Yolanda spelled $\frac{4}{5}$ of the words correctly. Sid spelled $\frac{7}{10}$ of the words correctly. Eric spelled $\frac{3}{4}$ of the words correctly. Ulysses spelled $\frac{17}{20}$ of the words correctly. Who spelled the most words correctly on the spelling test? By how much?

Show your work using words, numbers, and/or pictures.

[Student's handwritten work:]

Yolanda   Sid   Eric   Ulysses

Ulysses spelled most words correctly by 15 from Yolanda, 10 from Sid, 6 from Eric. Yolanda got the least words correctly spelled. 15+10+6=31

Who spelled the most words correctly on the spelling test? __Ulysses__

By how much? __31/20__

9 Janie's basketball team rents the school gym in the evening for some extra practice time. The school charges the $7.50 per hour ($h$) plus a basic charge of $5.00 for janitorial services.

Write an equation that represents the total charge, c, of renting the gym for h hours.

10 Marcus and his father are baking cookies for a school bake sale. They can make two dozen cookies every 30 minutes.

Draw a graph that represents the number of cookies they will have made at any point during the 3 hours they work.

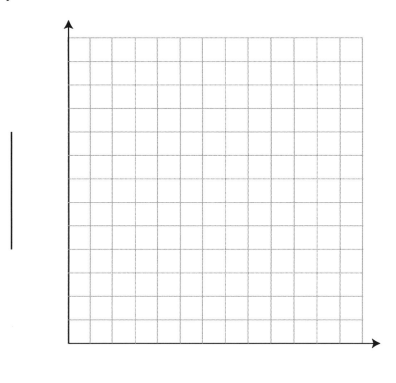

11 When Wei-Mah checked the weather forecast for Sunday, the predicted high temperature was 24°C. She remembered from Science class that the formula °F = $(\frac{9}{5})$ x °C + 32 should be used to convert Celsius to Fahrenheit.

What is the predicted high temperature for Sunday in degrees Fahrenheit?

○ A. 15.3° F

○ B. 75.2° F

○ C. 81.6° F

○ D. 100.8° F

12 A local bakery sells doughnuts by the dozen only. The first dozen costs $10.00. Each dozen after that costs $8.50. Mario is buying doughnuts for a fundraiser breakfast at his school. This information is displayed in the equation below, where *n* represents the number of dozens of doughnuts Mario purchased; $10.00 + (*n* – 1)($8.50).

How much money did Mario spend if he purchased 19 dozen doughnuts?

○ A. $153.00

○ B. $163.00

○ C. $238.00

○ D. $246.50

Go on▶

13 Marissa is tying bows to use as decorations for the holidays. She has 72 inches of ribbon, and each bow requires 2.4 inches of ribbon.

How many bows will Marissa be able to tie?

- ○ **A.** 24 bows
- ○ **B.** 28 bows
- ○ **C.** 30 bows
- ○ **D.** 36 bows

Go on▶

14 In Seattle, a taxi ride costs $1.80 per mile plus a $3.00 pick-up fee.

Write expressions to show how much it costs (c) to take a cab for 12 miles, 18 miles, and n miles.

How far can someone get in a cab for $48.00? Verify your solution.

12 miles:

18 miles:

n miles:

How far can someone get in a cab for $48.00? _____

Go on ▶

15 If a 250-piece puzzle has 70 border pieces, what is the **ratio** of border pieces to the total number of pieces in the puzzle?

A. $\frac{7}{25}$

B. $\frac{7}{18}$

C. $\frac{18}{25}$

D. $\frac{1}{7}$

Go on▶

16 Trisha drove her car 348 miles in 2 weeks and used 12 gallons of gasoline.

How many miles per gallon did Trisha's car get during these 2 weeks? Explain your answer.

How many miles per gallon did Trisha's car get during these 2 weeks? _____

17 Mike's laptop battery shows it has 30% of its power remaining.

What is 30% written as a decimal?

A. 0.030

B. 0.30

C. 3.0

D. 30.0

18 A jacket at Jeff's favorite store is advertised as being 20% off. The sale price is $64.

What was the original price of the jacket? Verify your solution.

Go on ▶

19 The diameter of a tire is about 2 feet.

How does the diameter compare to the circumference of the tire?

A. The circumference is equal to the diameter.

B. The circumference is about 2 times the diameter.

C. The circumference is about 3 times the diameter.

D. The circumference is about 6 times the diameter.

Go on▶

20 Kaitlyn is eating animal crackers. She wants to know the probability that any given animal cracker she takes from the box will be a lion. Selecting the crackers at random, she finds 4 lion crackers and 18 crackers that are not lions.

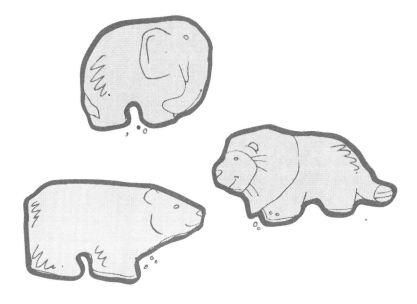

What is the **experimental probability** that a given cracker from the box is a lion?

A. $\frac{2}{11}$

B. $\frac{2}{7}$

C. $\frac{2}{9}$

D. $\frac{4}{18}$

Go on ▶

21 Phyllis has a box of cookies. There are 6 sugar cookies in the box, and there are 8 oatmeal cookies in the box.

If Phyllis takes a cookie from the box without looking, what is the **probability** she will get an oatmeal cookie?

A. $\frac{3}{7}$

B. $\frac{4}{7}$

C. $\frac{3}{4}$

D. $\frac{4}{3}$

Go on ▶

22 Mr. Gardner needs to put some fencing around his circular vegetable garden to keep out the rabbits. His garden is 12 feet in diameter.

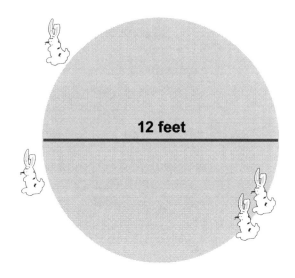

About how much fencing should he buy to enclose the garden?

A. 12 ft

B. 24 ft

C. 38 ft

D. 60 ft

Go on ▶

23 Annie is redecorating her bedroom. She will be adding new carpet and putting a wallpaper border around her entire room. To find out how much material she will need to buy, Annie must find the perimeter and area of her room.

Determine the **area** and **perimeter** of Annie's room below.

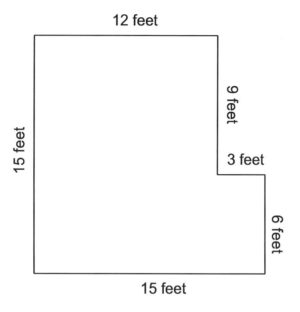

The perimeter of Annie's room:

The area of Annie's room:

Go on ▶

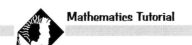

24 Jeffrey is building a large circular stage for an open-air play he is performing with his friends. He determined that the distance around the stage is about 44 feet.

What is the distance from the outer edge of the stage to the center where Jeffery has built a prop in the shape of a tree?

What is the area of the stage?

Show your work using words, numbers, and/or pictures.

What is the distance from the outer edge of the stage to the center? _____

What is the area of the stage? _____

Go on▶

25 Senay is making a mobile for his baby sister. He is using paper to cut out nets of three-dimensional figures and then folding them.

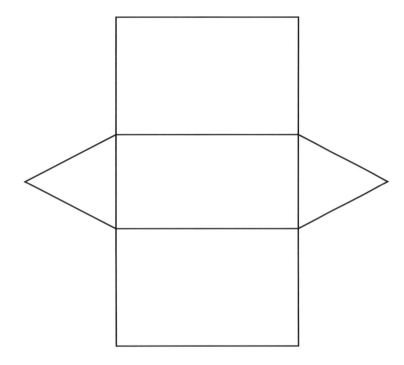

Which three-dimensional figure is represented by the net?

A. Triangular pyramid

B. Triangular prism

C. Rectangular pyramid

D. Rectangular prism

Go on ▶

26 Betsy is building a box with decorative cardboard.

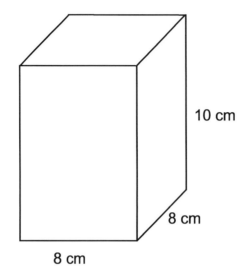

How much cardboard is needed to make the box? Explain your answer.

How much cardboard is needed to make the box? _____

What is the **volume** the box can hold? Explain your answer.

What is the volume the box can hold? _____

27 Ruby's father helped her build a square pyramid out of wood for her Social Studies class. Ruby wants to cover all sides of the pyramid with gold paper.

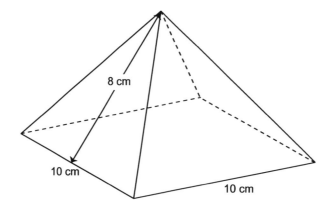

What is the **surface area** of the pyramid that Ruby will need to cover?

- A. 260 cm²
- B. 400 cm²
- C. 420 cm²
- D. 800 cm²

28 Juan is playing a trivia game with his friends. David asks Juan, "What type of polyhedron has four non-parallel triangular faces and one square face?" What should Juan answer?

- A. Triangular prism
- B. Square pyramid
- C. Triangular pyramid
- D. Rectangular prism

Go on▶

29 Ken wants to find the total number of hours he watched television this week.

Days	Sunday	Monday	Tuesday	Wednesday	Thursday	Friday	Saturday
Hours	$2\frac{1}{2}$	1	2	$2\frac{1}{2}$	$1\frac{1}{2}$	3	$4\frac{1}{2}$

Use Ken's chart to find the total.

○ **A.** 15 hours

○ **B.** $15\frac{1}{2}$ hours

○ **C.** $16\frac{1}{2}$ hours

○ **D.** 17 hours

Go on ▶

30 Paneet checked the thermometer at 8:00 a.m. The temperature fell 26° by 6:00 p.m.

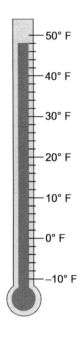

What is the temperature at 6:00 p.m.?

○ **A.** 22° F

○ **B.** 26° F

○ **C.** 32° F

○ **D.** 48° F

Go on▶

31 The temperatures for a four-day period in winter were measured in four different cities.

Which list correctly shows the temperatures ordered from **least** to **greatest**?

- A. 1° F, –13° F, –27° F, 35° F
- B. –5° F, –7° F, 18° F, –20° F
- C. –11° F, –9° F, 2° F, 12° F
- D. –15° F, –5° F, –7° F, 8° F

32 Rufus wants to paint the bike ramp below. One can of paint will cover 15 square feet.

What does Rufus need to know to calculate how much paint to buy?

- A. The volume of the ramp
- B. The cost to paint the ramp
- C. The surface area of the ramp
- D. How long it will take to paint the ramp

Go on ▶

33 Iris has a garden that is 12 feet by 15 feet. She wants to plant tulips. A bag of tulips covers 9 square feet of garden. Each bag costs $9.95. How many bags does she need to buy?

What information is **unnecessary** to solve this problem?

- A. The dimensions of the garden
- B. How many square feet a bag of tulips covers
- C. The area of the garden
- D. The cost of a bag of tulips

Go on▶

34 Curtis is planning to buy a new radio that costs $29.99, but he has a coupon for $5.00 off the lowest price of any item. When he gets to the store, the radio is on sale, marked down 10%. Sales tax is 6%.

What steps should Curtis take to find the total price of the radio?

What is the total price of the radio in dollars to the nearest cent?

Show your work using words, numbers, and/or pictures.

Steps Curtis should take to find the total price:

What is the total price of the radio in dollars to the nearest cent?

What is the total price of the radio in dollars to the nearest cent? _____

Go on ▶

35 On her walk to work, Bridget must walk $\frac{3}{4}$ mile from her car to the building. On Monday, she walked for 8 minutes and stopped $\frac{2}{3}$ of the way through her walk at the coffee shop.

How far had Bridget walked when she reached the coffee shop?

What percent of her planned total distance had Bridget completed when she stopped at the coffee shop?

Explain your answers.

Distance Bridget walked before stopping at the coffee shop:

Percent of distance completed when Bridget stopped at the coffee shop:

Go on ▶

If it took Bridget 8 minutes to walk to the point where she stopped, assuming she walked at the same pace and did not stop again, how long would it have taken Bridget to get to her work building from the coffee shop?

How long would it have taken Bridget to get to her
work building from the coffee shop? _____

Go on ▶

36 Estelle is putting new tile in her utility room. The width of a room is 12 feet. Tile comes in squares that are $\frac{3}{4}$ foot by $\frac{3}{4}$ foot.

How many tiles will fit along the wall?

How many tiles will fit along the wall?_____

Go on ▶

37 Daven knows that the diameter of a circle is twice the length of the radius. He notices that a tennis ball is a three-dimensional version of a circle, called a sphere. The radius of a tennis ball is 3.75 centimeters.

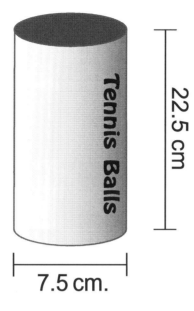

How many tennis balls will Daven be able to fit in this container?

- ○ **A.** 1 tennis ball
- ○ **B.** 2 tennis balls
- ○ **C.** 3 tennis balls
- ○ **D.** 6 tennis balls

Go on ▶

38 Mrs. Sato has a vegetable garden with 20 rows of vegetables. The garden has 6 rows of carrots. Each row of carrots has 15 carrots planted in it.

If Mrs. Sato sees a rabbit in her garden, what is the probability that the row the rabbit is in has carrots planted in it? Explain your answer.

| |
| |
| |
| |
| **What is the probability that the row the rabbit is in has carrots planted in it?** _____ |

39 Building contractors often estimate the cost of building a house based on its area in square feet. If a contractor charges $100 per square foot, a 1,200 square foot house has an estimated cost of $120,000 (1,200 x $100 = $120, 000). A customer asks for an estimate to build two circular houses.

If one house has a radius of 20 feet and the other house has a radius of 40 feet, which of the following statements is **true**?

○ A. The larger house will cost twice as much to build as the smaller house.

○ B. The larger house will cost three times as much to build as the smaller house.

○ C. The larger house will cost four times as much to build as the smaller house.

○ D. The larger house will cost eight times as much to build as the smaller house.

STOP

Mathematics Assessment

The Mathematics Assessment is made up of multiple-choice questions, completion items, short-answer questions, and extended-response questions. These questions show you how the skills you have learned in Mathematics class may be tested on the Mathematics MSP. The questions also give you a chance to practice your skills. If you have trouble with an area, talk with a parent or teacher.

Read each question carefully. If you do not know an answer, you may skip the question and come back to it later.

When you finish, check your answers.

Mathematics Assessment

Directions for Mathematics Assessment

Today you will take the Mathematics Assessment. On this part of the assessment, you are permitted to use tools such as calculators, rulers, protractors, or manipulatives.

Directions to the Student

There are several different types of questions on this assessment:

1. Some questions will ask you to choose the best answer from among four answer choices. These items are worth one point.

2. Some questions will ask you to write or draw an answer neatly and clearly **inside** an answer box.

    - Some of these questions are short. They may ask you to write an answer, to show how you got your answer using words, numbers, or pictures, or show the steps you used to solve the problem. These items are worth two points.

    - Others ask for more details (graphs, tables, written summaries). These questions also provide you with more room for your answer. These items are worth four points.

Here are some important things to remember as you take this assessment:

1. Read each question carefully and think about the answer.

2. When you choose a multiple-choice answer, make sure you completely fill in the circle next to the answer. Erase completely any marks that you want to change on multiple-choice items.

3. When an answer box is provided, write your answer neatly and clearly **inside** the box and show all your work. Cross out any work you do not want as part of your answer. **Do not use scratch paper.**

4. Use only a **No. 2 pencil**, not a mechanical pencil or pen, to write your answers. If you do not have a No. 2 pencil, ask your teacher to give you one.

5. You should have plenty of time to finish every question on the assessment. If you do not know the answer to a question, go on to the next question. You can come back to that question later.

6. When you reach the word **STOP** in your booklet, you have reached the end of the Assessment. Do **not** go on until you are told to turn the page.

7. If you finish early, you may check your work in this section **only**.

Go on▶

1. Barney is buying pizzas for a school party. He buys 12 pepperoni pizzas and 16 cheese pizzas. What is the ratio of pepperoni pizzas to cheese pizzas?

   ○ A. 2 to 3

   ○ B. 3 to 4

   ○ C. $\frac{3}{7}$

   ○ D. $\frac{6}{8}$

2. In Ted's freezer, there is a large box of freezer pops. The box contains 8 green freezer pops, 9 orange freezer pops, 10 red freezer pops, and 5 blue freezer pops.

   If one freezer pop is drawn randomly from the box, what is the probability it will be a color other than green?

   ○ A. $\frac{1}{4}$

   ○ B. $\frac{11}{16}$

   ○ C. $\frac{5}{8}$

   ○ D. $\frac{3}{4}$

Go on▶

3 The wheel of the largest unicycle in the world has a diameter of 73 inches.

How far would the unicycle travel with one revolution of the wheel? Explain your answer.

How far would the unicycle travel with one revolution of the wheel? _____

4 For a local art fair, Giuseppe drew a large mural on the street with sidewalk chalk. The dimensions of his mural were 5.4 feet by 7.25 feet.

How many square feet does Giuseppe's mural cover?

○ **A.** 17.28 square feet

○ **B.** 39.15 square feet

○ **C.** 51.30 square feet

○ **D.** 65.25 square feet

Go on ▶

5 Madison is testing light bulbs at the factory. She finds 12 light bulbs that don't work and 108 light bulbs that do work.

What is the **experimental probability** that a given light bulb at the factory doesn't work?

- A. $\frac{1}{12}$
- B. $\frac{1}{10}$
- C. $\frac{1}{9}$
- D. $\frac{12}{108}$

6 Jackson is refinishing his grandmother's old circular dining room table. The diameter of the table top is 6 feet.

What is the **area** of the table top that Jackson must refinish?

- A. 9.42 square feet
- B. 18.84 square feet
- C. 28.26 square feet
- D. 113.04 square feet

Go on▶

7 Chloe's new puppy eats $1\frac{1}{4}$ cups of food a day. Barry's dog eats $2\frac{1}{2}$ times as much food each day as Chloe's puppy eats.

How much food does Barry's dog eat each day?

○ A. $2\frac{1}{8}$ cups

○ B. $3\frac{1}{8}$ cups

○ C. $3\frac{1}{4}$ cups

○ D. $3\frac{3}{4}$ cups

8 What type of polyhedron has 6 faces and 8 vertices?

○ A. Rectangular prism

○ B. Triangular prism

○ C. Rectangular pyramid

○ D. Triangular pyramid

Go on ▶

9 Pacey, Dawson, and Joey are thinking about running a relay race together. Each leg of the race is $\frac{1}{4}$ of a mile. Pacey can run $\frac{1}{4}$ mile in 1.93 minutes. Dawson can run $\frac{1}{4}$ mile in 1.85 minutes. Joey can run $\frac{1}{4}$ mile in $2\frac{1}{4}$ minutes.

What is the average amount of time it takes each runner to run $\frac{1}{4}$ mile?

> **What is the average amount of time it takes each runner to run $\frac{1}{4}$ mile?** _____

Go on▶

10 Marty buys coffee from the local coffee shop every day. He wants to calculate how much money he spends (*c*) on coffee. Each cup of coffee costs $1.85.

Write expressions to show how much money Marty spends on coffee in 7 days, 15 days, and *n* days.

How many days did it take Marty to spend $111.00 on coffee?

Show your work using words, numbers, and/or pictures.

Amount Marty spends on coffee in 7 days:

Amount Marty spends on coffee in 15 days:

Amount Marty spends on coffee in *n* days:

Number of days until Marty spends $111.00 on coffee:

Go on ▶

11. Kyle is surveying his apartment before he gets it ready to sell.

What is the **perimeter** of Kyle's apartment?

What is the **area** of Kyle's apartment?

Show your work using words, numbers, and/or pictures.

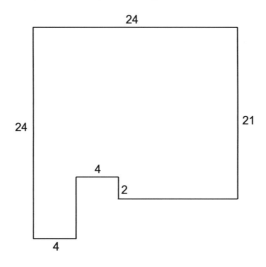

What is the perimeter of Kyle's apartment? _____

What is the area of Kyle's apartment? _____

12 The local museum is displaying an exhibit on ancient Egypt. Included in the exhibit is a model of a pyramid made from foam sheets.

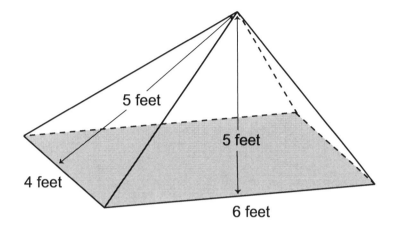

What is the **area** of foam sheets that were used to create the pyramid, including the base?

○ A. 74 square feet

○ B. 100 square feet

○ C. 124 square feet

○ D. 600 square feet

Go on ▶

13 Renuka makes pizzas at the local pizzeria as an after-school job. She can make 5 pizzas every 20 minutes.

Draw a graph that represents the number of pizzas Renuka will have made at any point during the 2 hours she works.

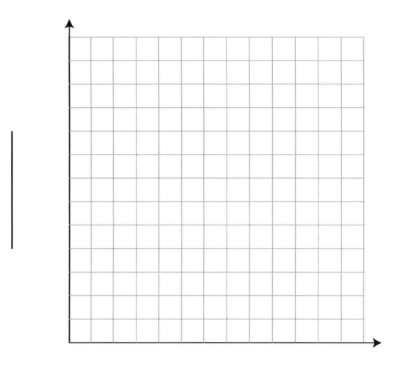

14 Mitch wants to find the total number of hours he slept last week.

Days	Sunday	Monday	Tuesday	Wednesday	Thursday	Friday	Saturday
Hours	$7\frac{1}{2}$	$8\frac{1}{4}$	7	$6\frac{3}{4}$	8	$9\frac{1}{4}$	9

Use Mitch's chart to find the total.

○ A.   54 hours

○ B. $55\frac{3}{4}$ hours

○ C.   56 hours

○ D. $56\frac{1}{4}$ hours

15 Sarah is measuring the temperature of a mystery liquid as it heats up. She records the temperature at 30-second intervals, but forgets to write them in order. The temperatures are 67° F, –15° F, 14° F, 35 °F, –2° F.

Which list correctly shows the temperatures ordered from **coldest** to **warmest**?

○ A.   –2° F, 14° F, –15° F, 35° F, 67° F

○ B.   –2° F, –15° F, 14° F, 35° F, 67° F

○ C.   –15° F, –2° F, 14° F, 35° F, 67° F

○ D.   67° F, 35° F, 14° F, -2° F, -15° F

Go on ▶

16 Ruby and Lexie are buying candy at the store. Ruby fills her bag $\frac{1}{3}$ of the way full. Lexie fills hers $\frac{1}{2}$ as full as Ruby's.

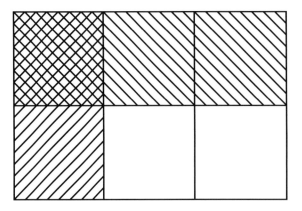

Which equation is pictured in the model and can be used to calculate how full Lexie's bag is?

○ A. $\frac{1}{3} \div \frac{1}{2}$

○ B. $\frac{1}{3} \times \frac{1}{2}$

○ C. $\frac{2}{3} \times \frac{1}{2}$

○ D. $\frac{1}{3} + \frac{1}{2}$

Go on ▶

17 Dana is buying a new home. Her realtor told Dana that the rectangular backyard is 702 square feet. Dana already knows that the backyard is the same width of her house, which is 33.75 feet wide.

About how deep is Dana's backyard?

○ A.   18 feet

○ B.   20 feet

○ C.   20.8 feet

○ D.   25 feet

18 Kurt's savings account has an annual interest rate of 1%. So each year he receives a credit to his account equal to 1% of his balance. At the end of 2008, Kurt's balance was $5,320.

How much money did Kurt make in interest in 2008?

○ A. $0.53

○ B. $5.32

○ C. $53.20

○ D. $532.00

Go on▶

19 A recycling center pays $0.06 per pound, up to 100 pounds, for emptied aluminum cans. After the first 100 pounds, the rate increases to $0.10 per pound. The center uses the following expression to calculate how much to pay someone who drops off more than 100 pounds of emptied aluminum cans: $6.00 + $0.10(n − 100).

How much would the center pay someone who dropped off 165 pounds of emptied aluminum cans?

- ○ A. $6.65
- ○ B. $12.50
- ○ C. $396.50
- ○ D. $906.50

20 The wheel of Ann's bike spun around 95 times in 15 seconds.

At what rate was Ann's bike wheel spinning? Explain your answer.

At what rate was Ann's bike wheel spinning? _____

Go on ▶

21 Bruce is trying to calculate the cost of each mile per gallon on his 2009 car. He put 14.926 gallons into his tank that cost $2.12 per gallon. Bruce drove 348 miles before he had to fill up his tank again.

What information is **unnecessary** to find out the cost of each mile per gallon?

- ○ A. The number of miles he drove
- ○ B. The number of gallons he put in his tank
- ○ C. The cost per gallon of gasoline
- ○ D. The year the car was made

22 Robby's sister must mail out 120 wedding invitations. Each invitation will require $0.60 postage. Stamps are available in the following denominations: $0.01, $0.05, $0.19, $0.39, and $0.45.

What question can be answered using the information given?

- ○ A. What is the least amount of stamps Robby's sister can buy and still mail all the invitations?
- ○ B. How much change will Robby's sister receive at the post office when she buys the stamps?
- ○ C. How long will it take Robby's sister to buy the stamps at the post office?
- ○ D. What is the largest size envelope that can be mailed for $0.60?

Go on ▶

23 Charlie is looking at a net.

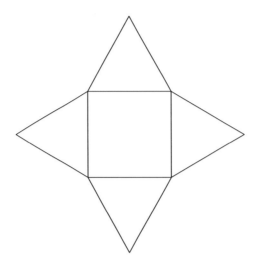

Which **three-dimensional figure** is represented by the net?

○ **A.** Triangular pyramid

○ **B.** Triangular prism

○ **C.** Square pyramid

○ **D.** Rectangular prism

24 The Upstart Communications Company offers customers a special long distance calling rate that includes a $0.25 connection fee applied to each long distance call plus a $0.10 per minute charge.

Which equation represents the cost of a call, c, that is a number of minutes, m, long?

○ **A.** $c = \$0.35m$

○ **B.** $c = \$0.25m + \$0.10$

○ **C.** $c = \$0.25 + \$0.10m$

○ **D.** $c = \$0.25m + \$0.10m$

25 Madeleine has to read a 350-page book in two weeks. She reads the same number of pages every night.

How many pages does she read each night?

○ **A.** 14 pages

○ **B.** 25 pages

○ **C.** 35 pages

○ **D.** 175 pages

Go on ▶

26 Mary Jo drove 200 miles to see her family for the holidays. On the way there, she drove an average speed of 60 miles per hour but on the way home, she averaged only 50 miles per hour.

How much longer did the return trip take than the original trip? Explain your reasoning.

Show your work using words, numbers, and/or pictures.

How much longer did the return trip take than the original trip? _____

Go on ▶